Chaos Bound

ALSO BY N. KATHERINE HAYLES

The Cosmic Web: Scientific Field Models and Literary Strategies in the Twentieth Century

Chaos Bound

ORDERLY DISORDER IN
CONTEMPORARY LITERATURE
AND SCIENCE

N. Katherine Hayles

CORNELL UNIVERSITY PRESS

ITHACA AND LONDON

First published 1990 by Cornell University Press.

International Standard Book Number (cloth) 0-8014-2262-0
International Standard Book Number (paper) 0-8014-9701-9

Library of Congress Catalog Card Number 89-23893
Printed in the United States of America

Librarians: Library of Congress cataloging information appears on the last page of the book.

♾ The paper used in this publication meets the minimum requirements of the American National Standard for Permanence of Paper for Printed Library Materials Z39.48–1984.

FOR MY FAMILY, ESPECIALLY
MY PARENTS, THELMA AND EDWARD BRUNS,
AND MY CHILDREN, LORI AND JONATHAN

Contents

Illustrations

Preface

SOME years ago I began to wonder why different disciplines, sufficiently distant from one another so that direct influence seems unlikely, should nevertheless focus on similar kinds of problems about the same time and base their formulations on isomorphic assumptions. It seemed to me then, and still does seem, that the most plausible explanation is cultural. Different disciplines are drawn to similar problems because the concerns underlying them are highly charged within a prevailing cultural context. Moreover, different disciplines base the theories they construct on similar presuppositions because these are the assumptions that guide the constitution of knowledge in a given episteme. This position implies, of course, that scientific theories and models are culturally conditioned, partaking of and rooted in assumptions that can be found at multiple sites throughout the culture.

My first attempt to demonstrate this phenomenon focused on the field concept that was central to several disciplines during the first half of the century. The field concept implied that reality consists not of discrete objects located in space but rather of an underlying field whose interactions *produce* both objects and space. It further implied (and this was perhaps its most important consequence for literature) that there is no exterior, objective viewpoint from which to observe, for one is always already within the field, caught in and constituted through the very interactions that one is trying to describe. Field models were associated with the emergence of inherent limits on what can be articulated. The realization that these limits were inescapable was decisive, for it dealt a death blow to foundationalist movements in physics, mathematics, and, in a rather different way, literary theory. In *The Cosmic Web: Scientific Field Models and Literary Strategies in the Twentieth Century*, I traced the scientific development of the field concept through the special and general theories of relativity, quantum mechanics, and the foundations of mathematics, especially Gödel's theorem, the Church-Turing theorem, and the Halting problem. Corresponding literary strategies were explored and illustrated through such writers as D. H. Lawrence, Vladimir Nabokov, Robert Pirsig, and Thomas Pynchon.

In the years since *The Cosmic Web* was published, it has become increasingly clear that the developments explored there were part of a continuing sequence of events. Painted in broad strokes, the picture looks something like this. During the first half of the century, many disciplines were preoccupied by attempts to develop totalizing theories that could establish unambiguous connections between theory and observation, articulation and reality. By mid-century, virtually all of these attempts had been defeated or had undergone substantial modification. In mathematics, Gödel's theorem led to a crisis in the foundations of mathematics which still is not resolved; in physics, relativity theory was combined with quantum mechanics to form quantum field theory; within quantum mechanics itself, the uncertainty principle gave rise to such variants as Bell's theorem, the many worlds theory, and hidden variables. Literary studies were a relative stronghold of totalizing theories by comparison, for archetypal criticism and structuralism sought to find in mythemes or other underlying structural elements the foundations that had already by 1950 been largely repudiated in the physical sciences.

Then the pendulum, having gone as far as it could in the direction of encompassing order, began to swing the other way as various disciplines became interested in exploring the possibilities of disorder. Attention focused on the mechanisms that made unpredictability a fact of life rather than the aberration it had seemed in Newtonian mechanics. In physics, complex systems were at the center of research in nonlinear dynamics, fluid mechanics, and quantum electrodynamics. In mathematics, fractal geometry burst upon the scene. In thermodynamics, important results were derived from irreversible systems far from equilibrium. In biology, systems theory implied that disorder at one level of communication within an organism would become order on another. Even in such traditionally statistical fields as meteorology and epidemiology, new ways of thinking about erratic variations were revealing deep structures of order within the apparent disorder.

The shift in literary theory from structuralism to poststructuralism is too well known to need further commentary here. A characteristic turn was marked in 1967 by Roland Barthes's essay "From Science to Literature" in the *Times Literary Supplement*, when he announced that structuralism would fail unless it was able "to call into question the very language by which it knows language." With the embedding of language in itself, reflexivity became unavoidable and, with the emergence of deconstruction, was linked to an inability to determine or establish origins. The radical instabilities that were then produced within texts led to an interest in disorder and unpredictability in literature analogous to that in the sciences. The turn toward disorder was paralleled in contemporary fiction by the emergence of postmodernism, exemplified by the works of such writers as William Gaddis, Don De Lillo, Robert Coover, and William Burroughs.

Chaos Bound traces these developments in literature and science and locates them within postmodern culture, particularly within the technologies and social landscapes created by the concept of information. *Chaos Bound* is both sequel and complement to *The Cosmic Web*. The paradigm of orderly disorder may well prove to be as important to the second half of the century as the field concept was to the first half. Moreover, it brings into focus aspects of complex systems which had begun to surface within field models but which could not be adequately accounted for by them. In this sense, *Chaos Bound* both documents what happened after the events described in *The Cosmic Web* and concentrates on what had been left out of the earlier models.

Though there are thus important connections between the two books, there are also substantial differences. Especially notable is the increased emphasis in *Chaos Bound* on locating science and literature within contemporary culture. The recurrent image I use to explain the complex interconnections of theory, technology, and culture is a feedback loop. The increased mobility of troops and supplies in World War II, for example, made accurate information a much more important factor in military strategy than it had ever been before. Consequently, there was intensive research on the concept of information. In the years immediately following the war, the theories that emerged from this research were translated into new technologies, which in turn transformed the cultures of highly developed countries in ways both subtle and profound. These transformations stimulated the creation of new methods of analysis for complex systems, for society itself had become a complex system in a technical sense. Thus the feedback cycle connected theory with culture and culture with theory through the medium of technology. Literary texts and theories were also involved in this cycle, for they too were affected by technology at the same time that they were affecting it. It should be no surprise, then, that many of the presuppositions that underlie the literary texts are also embedded within the scientific models and theories of the period.

These similarities notwithstanding, different disciplinary traditions can impute strikingly different values to isomorphic paradigms. In the physical sciences, for example, nonlinear dynamics is seen as a way to bring complex behavior within the scope of rational analysis. Analogous theories in literary studies, by contrast, are often embraced because they are seen as resisting totalizing theories. This double edge to the current preoccupation with chaos—the ambiguity of whether it brings chaos within rational compass or signals the final defeat of totalizing projects—suggests that disciplinary traditions can play crucial roles in determining how isomorphic ideas are valued and interpreted. It also suggests that postmodern culture authorizes both of these visions. In the final chapter I argue that this divided impulse is in fact deeply characteristic of postmodern culture. Concerned to resist totalization, postmodern theories image and enact a totalization more complete than any that came before.

Many colleagues and friends have given invaluable help over the years I have been working on this book. Adalaide Morris, Mary Lou Emery, Herman Rapaport, Elizabeth Ermarth, John Nelson, Wayne

Polyzou, Herbert Hethcote, and Stephen Brush read portions of the manuscript and offered suggestions for revisions. The Project on the Rhetoric of Inquiry devoted a seminar to discussion of a chapter, from which I gleaned many helpful suggestions. The Society for the Humanities at Cornell University provided a forum for the presentation of another chapter. The Woodrow Wilson International Center for Scholars, through a Woodrow Wilson Fellowship and the use of its facilities, provided support for an early phase of the project, as did the University of Iowa through a Faculty Fellowship for much of the later work. University House at the University of Iowa provided research facilities, intellectual nurturance, and solitude when they were much needed. The Humanities Division of California Institute of Technology provided research support in the final stages of the writing. The statements and views expressed in this book are of course solely my responsibility and not necessarily those of any of the people or groups I have named.

I am grateful to the presses and editors who generously gave permission to use material that has previously appeared. Chapter 2 appeared in slightly altered form in *Literature and Science: Theory and Practice*, edited by Stuart Peterfreund, copyright © 1990 by Stuart Peterfreund, and is used by permission of Northeastern University Press. A portion of chapter 7 appeared in *SubStance* and is used by permission of the University of Wisconsin Press. Another part of chapter 7 appeared in *One Culture: Essays in Science and Literature*, edited by George Levine, copyright © 1987 by the Board of Regents of the University of Wisconsin System, and is used by permission of the University of Wisconsin Press. Chapter 5 appeared in *Science-Fiction Studies* and portions of chapters 7 and 8 appeared in *New Literary History*; they are used by permission.

I am pleased to acknowledge the contributions of Todd Erickson, who supplied figures 1 and 2; Nural Akchurin, who supplied figure 9; and Herbert Hethcote, who supplied figure 4. I thank James Crutchfield and *Scientific American* for permission to reproduce figures 3 and 5, and W. H. Freeman Company for permission to use figures 7 and 8.

I owe a large debt to Jules van Lieshout, who patiently worked to bring consistency and elegance to the manuscript, and to bring the manuscript to the computer. Without his help, it would no doubt have taken several additional months to complete this work. I owe thanks to Zofia Lesinska and, finally, to my students past and present, for I rarely left a class without having learned from them nearly as much as I taught. My greatest debt is to my family and friends, who met late dinners, postponed engagements, and lost weekends with understanding and love.

N. KATHERINE HAYLES

Iowa City, Iowa

Chaos Bound

CHAPTER 1

Introduction: The Evolution of Chaos

> I am an old man now, and when I die and go to Heaven there are two matters on which I hope for enlightenment. One is quantum electrodynamics, and the other is the turbulent motion of fluids. And about the former I am really rather optimistic.
>
> —Sir Horace Lamb, 1932

It all started with the moon. If only the earth could have gone round the sun by itself, unperturbed by the complications in its orbit which the moon's gravitational field introduced, Newton's equations of motion would have worked fine. But when the moon entered the picture, the situation became too complex for simple dynamics to handle. The moon attracted the earth, causing perturbations in the earth's orbit which changed the earth's distance from the sun, which in turn altered the moon's orbit around the earth, which meant that the original basis for the calculations had changed and one had to start over from the beginning. The problem was sufficiently complex and interesting to merit a name and a prize of its own. It became known as the three-body problem, and the king of Sweden offered a reward to the first person who could prove a solution was possible.[1] Instead, in 1890 Henri Poincaré (whose formidable talent was responsible for creating topology and half a dozen other new fields) published a paper proving that in general, a solution was not possible by means of Newtonian equations (Poincaré,

[1] I am indebted to Phillip Holmes for drawing this incident to my attention.

1890). With commendable foresight, the king of Sweden gave Poincaré the prize anyway. Perhaps he intuited that Poincaré's work had opened a window on a kind of world that Newtonian mechanics had not envisioned. By proving that the introduction of small perturbations into linear equations was not in general sufficient to solve nonlinear problems, Poincaré implied that a new kind of science and mathematics was necessary to account for the dynamics of complex systems. From this realization the science of chaos was born.

But not immediately. Positivism was in full swing throughout Western Europe and America, and mathematicians were preoccupied by efforts to put mathematics on a firm foundation by formalizing it. By 1931, when Kurt Gödel dashed these hopes by proving that formal systems could not be axiomatized completely (Gödel, 1962), the clue that Poincaré's work provided for the labyrinthine difficulties of complex dynamics was in danger of dropping out of sight. (An exception was Russia, where many important results were obtained during the 1930s and 1940s.) In the West, the study of complex dynamics did not come into its own until computers became widespread and readily accessible during the 1960s and 1970s.

The same two decades saw a significant intellectual shift throughout the human sciences. Its essence was a break away from universalizing, totalizing perspectives and a move toward local, fractured systems and modes of analysis. Just as new methods were being developed within the physical sciences to cope with the complexities of nonlinear systems, so new ways of reading and writing about literature were coming to the fore in critical theory. The (old) New Critics had taken for granted that a literary work was a verbal object, bounded and finite, however ambiguous it might be within. But the (new) New Critics saw textual boundaries as arbitrary constructions whose configurations depended on who was reading, and why. As books became texts, they were transformed from ordered sets of words to permeable membranes through which flowed the currents of history, language, and culture. Always already lacking a ground for their systems of signification, texts were not deterministic or predictable. Instead they were capable of becoming unstable whenever the slightest perturbation was introduced. The well-wrought urn, it seemed, was actually a reservoir of chaos.

Each of these developments appeared within a well-defined disciplinary tradition, and each is explicable in terms of what had preceded it within the discipline. Nonlinear dynamics, for example, traces its linage through Mitchell Feigenbaum and Edward Lorenz back to Poincaré; poststructuralism through such theorists as Jacques Derrida and Paul de Man back to Friedrich Nietzsche and Martin Heidegger. From the specialist's point of view, there is no need to go outside these boundaries to understand what happened. But there are also suggestive similarities across disciplinary lines. Suppose an island breaks through the surface of the water, then another and another, until the sea is dotted with islands. Each has its own ecology, terrain, and morphology. One can recognize these distinctions and at the same time wonder whether they are all part of an emerging mountain range, connected both through substrata they share and through the larger forces that brought them into being.

In this book, I argue that certain areas within the culture form what might be called an archipelago of chaos. The connecting theme is a shift in the way chaos is seen; the crucial turn comes when chaos is envisioned not as an absence or void but as a positive force in its own right. This is a three-sided study, triangulating among chaos theory, poststructuralism, and contemporary fiction. Also of concern is the cultural matrix from which all three sides emerge and with which they interact. Concerned with the physical sciences as well as literature, the study investigates language's power to constitute reality, and reality's power to constrain and direct language. It speculates about the broader cultural conditions that authorize the new visions of chaos, and inquires into how these conditions shape and are shaped by modern narratives.

The metaphor of the triangle implies, of course, that there are connections and relationships among the three sides. One of the challenges in literature and science is to develop methodologies that can illuminate convergences between disciplines, while still acknowledging the very real differences that exist. In my view, analogies between literary and scientific versions of chaos are important *both for the similarities they suggest and for the dissimilarities they reveal.* The similarities arose because of broadly based movements within the culture which made the deep assumptions underlying the new paradigms thinkable, perhaps inevitable, thoughts. They illustrate how feedback loops among theory, technology, and culture develop and expand into complex connections between literature and science which are mediated through the cultural matrix. The dissimilarities, by contrast, point to the importance of disciplinary traditions in guiding inquiry and shaping thought. To account for them, it is necessary to understand how and why certain questions became important in various disciplines before the appearance of the new paradigms. The dual emphasis on cultural fields and disciplinary sites implies a universe of discourse that is at once fragmented and unified. Cultural fields bespeak the interconnectedness of a world in which instantaneous global communication is a mundane reality; local differences acknowledge the power of specialization within contemporary organizations of knowledge.

The connections I explore among contemporary literature, critical theory, and science are not generally explainable by direct influence. Rather, they derive from the fact that writers, critics, and scientists, however specialized or esoteric their work, all share certain kinds of everyday experiences. Consider the following question: Why should John Cage become interested in experimenting with stochastic variations in music about the same time that Roland Barthes was extolling the virtues of noisy interpretations of literature and Edward Lorenz was noticing the effect of small uncertainties on the nonlinear equations that described weather formations? An influence argument would look for direct connections to explain these convergences. Sometimes such connections exist. It is possible that Barthes listened to Cage, Cage studied Lorenz, Lorenz read Barthes. But it stands to reason that, of all the interdisciplinary parallels one might notice, only a few will be connected by direct lines of influence, which are usually conveyed through *disciplinary* traditions. One could, for example, trace a clear line of descent from John Cage to Brian Eno to the Talking Heads and U2.

Interdisciplinary parallels commonly operate according to a different dynamic. Here influence spreads out through a diffuse network of everyday experiences that range from reading *The New York Times* to using bank cards on automatic teller machines to watching MTV. When enough of the implications in these activities point in the same direction, they create a cultural field within which certain questions or concepts become highly charged. Perhaps, for example,

Brian Eno might first learn about Roland Barthes through *Time* magazine. Intrigued, he might read one of Barthes's books. Or he might not. The brief article summarizing Barthes's ideas would then become one of the elements in Eno's cultural field, available to be reinforced by other elements until a resonance built up which was strong enough to be a contributing factor in his work.

Between 1960 and 1980, cultural fields were configured so as to energize questions about how stochastic variations in complex systems affected systemic evolution and stability. It is easy to see how the political movements of the 1960s contributed to this interest. Also important was the growing realization that the world itself had become (or already was) a complex system economically, technologically, environmentally. Along with the information capabilities of modern communication systems came the awareness that small fluctuations on the microscale could, under appropriate conditions, quickly propagate through the system, resulting in large-scale instabilities or reorganizations. A revolution in the Middle East, for example, could trigger a precipitous rise in oil prices, leading to energy shortages and inflationary spirals in the developed countries, which in turn could spark a global recession that would force major restructurings in international finance. When such cascading scenarios are ever-present possibilities, the realization that small causes can lead to very large effects is never far from consciousness. The ecological movement is a case in point. People concerned about the global environment are intensely aware that a seemingly small event—an inattentive helmsman on the bridge of an oil tanker, say—can have immediate and large-scale effects on an entire coastal area. Implicit in this awareness is increased attention to random fluctuations, and consequently to the role that chaos plays in the evolution of complex systems.

Another factor that helped to energize the concepts underlying the new paradigms was the realization that as systems became more complex and encompassing, they could also become more oppressive. In more than one sense, the Cold War brought totalitarianism home to Americans. As information networks expanded and data banks interlocked with one another, the new technology promised a level of control never before possible. In this paranoiac atmosphere, chaotic fluctuations take on an ambiguous value. From one point of view they threaten the stability of the system. From another, they offer the liberating possibility that one may escape the informational net by slipping along its interstices. In *Gravity's Rainbow*, for example, chaos reigns supreme in the "Zone," the free-floating, anarchical space that was Western Europe for a brief time at the end of World War II. Threatening as the Zone sometimes is, its chaotic multivalency marks the distance between Pynchon's postmodern text and the nightmare vision of Orwell's *1984*.

In the assigning of a positive value to chaos, information theories and technologies played central roles. In addition to creating the necessary technological landscape, they laid the theoretical foundation for conceptualizing chaos as a presence rather than an absence. Later chapters will explore this transformation, showing how a crucial move in the transvaluation of chaos was the separation of information from meaning. Once this distinction was made, the way was open for information to be defined as a mathematical function that depended solely on the distribution of message elements, independent of whether the message had any meaning for a receiver. And this step in turn made it possible to see chaotic systems as rich in information rather than poor in order.

Suppose I send a message that contains the series 2, 4, 6, 8 . . . and ask you to continue the sequence. Because you grasp the underlying pattern, you can expand the series indefinitely even though only a few numbers are specified. Information that has a pattern can be compressed into more compact form. I could have sent the message as "Enumerate the even integers, starting with 2." Or even more concisely, "Count by twos." By contrast, suppose I send you the output of a random number generator. No matter how many numbers I transmit, you will be unable to continue the sequence on your own. Every number comes as a surprise; every number conveys new information. By this reasoning, the more random or chaotic a message is, the more information it contains.

You may object that although the numbers are always new and surprising, they do not *mean* anything. The objection illustrates why it was necessary to separate information from meaning if chaotic systems were to be considered rich in information. Implicit in the transvaluation of chaos is the assumption that the production of information is good in itself, independent of what it means. Having opened this possibility by creating a formal theoretical framework that implied it, information and communication technologies actualized it in everyday life. Every time we keep a TV or radio going in the background, even though we are not really listening to it, we are acting out a behavior that helps to reinforce and deepen the attitudes that underwrite a positive view of chaos.

Stanislaw Lem in *The Cyberiad* has a fable that speaks to this point (Lem, 1974b). Two constructors, Trurl and Klapaucius, take a journey that brings them into the clutches of Pugg the PHT Pirate. Actually the name (which they know only from rumor) is a slight error. Pugg has a Ph.D, and what he craves above all else, even more than gold and obeisance from his subjects, is information. So Trurl and Klapaucius create for him a Demon of the Second Kind.

The First Kind of Demon (about which we will hear more later) was proposed by James Clerk Maxwell in 1859. To test the second law of thermodynamics, Maxwell imagined a mythical imp who presided over a box of ideal gas divided by a partition. The Demon's task was to sort the molecules by opening and closing a shutter in the partition, allowing only the fast molecules to pass through. The resulting separation created a temperature differential, which in turn could be converted into work. Lem refers to this history by allusion to its difference from the Demon of the Second Kind, who is an up-to-date version appropriate to an information age. Like his predecessor, the Second Demon also presides over a box of stale air. Instead of sorting the molecules, however, he watches their endless dance. Whenever the molecules form words that make sense, he writes them down with a tiny diamond-tipped pen on a paper tape. Whereas the First Demon uses randomness to produce work, the Second Demon uses it to produce information.

Pugg is delighted with the invention and immediately sits down to read the tape with his hundreds of eyes. He learns "how exactly Harlebardonian wrigglers wriggle, and that the daughter of King Petrolius of Labondia is named Humpinella, and what Frederick the Second, one of the paleface kings, had for lunch before he declared war against the Gwendoliths, and how many electron shells an atom of thermionolium would have, if such an element existed, and what is the cloacal diameter of a small bird called the tufted twit" (p. 157). As the list continues and the tape rolls on, Pugg is buried under its toils. The narrator informs us that he sits there to this day, learning "no end of things about rickshaws, rents and roaches, and about his own fate, which has been related here, for that too is included in some section of the tape—as are the histories, accounts and prophecies of all things in creation, up until the day the sun burns out; and there is no hope for him . . . unless of course the tape runs out, for lack of paper" (pp. 159–160).

The fable is at least as compelling today as it was when it was written in 1967. Like Pugg, we are increasingly aware that information is a commodity every bit as valuable as diamonds and gold. Indeed, it can often be converted directly into money (as recent insider trading scandals have demonstrated). What are the computer programs that large investment firms use for stock trading but Demons of the Second Kind? From random fluctuations in the market they extract information and money, thus justifying Maxwell's intuition that the second law of thermodynamics may have left something important out of account. Whether this project will succeed in the long run or bury us underneath it, as Pugg was entombed by his information, remains to be seen. Not in doubt is the important role that such phenomena play in reinforcing the connection between information and randomness. The more chaotic a system is, the more information it produces. This perception is at the heart of the transvaluation of chaos, for it enables chaos to be conceived as an inexhaustible ocean of information rather than as a void signifying absence.

Once the link was forged between chaos and information, a chain of consequences followed. To introduce them, I want to explain more about the structure and content of chaos theory. First, a disclaimer: "chaos theory" and the "science of chaos" are not phrases usually employed by researchers who work in these fields. They prefer to designate their area as nonlinear dynamics, dynamical systems theory, or, more modestly yet, dynamical systems methods. To them, using "chaos theory" or the "science of chaos" signals that one is a dilettante rather than an expert. Nevertheless, I will use these terms throughout my discussion, because part of my project is to explore what happens when a word such as "chaos," invested with a rich tradition of mythic and literary significance, is appropriated by the sciences and given a more specialized meaning. The older resonances do not disappear. They linger on, creating an aura of mystery and excitement that even the more conservative investigators into dynamical systems methods find hard to resist (especially when they apply for grants or explain their work to the public). As new meanings compete with traditional understandings within the sign of chaos, "chaos" becomes a highly charged signifier, attracting interest from many areas within the culture. The underlying forces that have fueled the new paradigms—the rapid development of information technologies, the increasing awareness of global complexities, and consequent attention to small fluctuations—do not depend on any single factor, especially one so slight as the choice of a name for the new theories. But the name *is* important, for in its multiple meanings it serves as a crossroads at which diverse paths within the culture meet.

Chaos theory is a wide-ranging interdisciplinary research front that includes work in such fields as nonlinear dynamics, irreversible thermodynamics, meteorology, and epidemiology. It can be generally understood as the study of complex systems, in which the nonlinear problems that perplexed Poincaré's contemporaries are considered in their own right, rather than as inconvenient deviations from linearity. Within chaos theory, two general emphases exist. In the first, chaos is seen as order's precursor and partner, rather than as its opposite. The focus here is on the spontaneous emergence of self-organization from chaos; or, in the parlance of the field, on the dissipative structures that arise in systems far from equilibrium, where entropy production is high. The realization that entropy-rich systems facilitate rather than impede self-organization was an important turning point in the contemporary reevaluation of chaos. A central figure in this research is Ilya Prigogine, who in 1977 won the Nobel Prize for his work with irreversible thermodynamics. The title of the book he co-authored with Isabelle Stengers, *Order out of Chaos*, provides the motto for this branch of chaos theory.

The second branch emphasizes the hidden order that exists *within* chaotic systems. Chaos in this usage is distinct from true randomness, because it can be shown to contain deeply encoded structures called "strange attractors." Whereas truly random systems show no discernible pattern when they are mapped into phase space, chaotic systems contract to a confined region and trace complex patterns within it. The discovery that chaos possesses deep structures of order is all the more remarkable because of the wide range of systems that demonstrate this behavior. They range from lynx fur returns to outbreaks of measles epidemics, from the rise and fall of the Nile River to eye movements in schizophrenics. Researchers associated with this branch of chaos theory include Edward Lorenz, Mitchell Feigenbaum, Benoit Mandelbrot, and Robert Shaw. The strange-attractor branch differs from the order-out-of-chaos paradigm in its attention to systems that remain chaotic. For them the focus is on the orderly descent into chaos rather than on the organized structures that emerge from chaos.

For a variety of reasons, fewer connections have been forged between the two branches than one might expect. The two branches employ different mathematical techniques to analyze chaos. Although some translations have been made, the different modes of analysis make communication between the branches difficult. There are also different views on what the research signifies. Prigogine has strong ties with French intellectual circles, and the order-out-of-chaos branch is known for its willingness to extrapolate beyond experimental results to philosophical implications. It has been criticized within the scientific community for the relative paucity of its results, especially in light of the large philosophical claims made for them. The strange-attractor branch, by contrast, has been if anything undertheorized; its practioners prefer to concentrate on problems of immediate practical interest. In brief, the order-out-of-chaos branch has more philosophy than results, the strange-attractor branch more results than philosophy.

These different orientations lead to different kinds of conclusions. Prigogine sees the primary importance of the order-out-of-chaos branch in its ability to resolve a long-standing metaphysical problem: it reconciles being with becoming. For him, chaos theory is revolutionary because of what it can tell us about the arrow of time. By comparison, the strange-attractor branch emphasizes the ability of chaotic systems to generate new information. Almost but not quite repeating themselves, chaotic systems generate patterns of extreme complexity, in which areas of symmetry are intermixed with asymmetry down through all scales of magnification. For researchers in this branch, the important conclusion is that nature, too complex to fit into the Procrustean bed of linear dynamics, can renew itself precisely because it is rich in disorder and surprise.

Perhaps because of these differences, James Gleick, in his influential narrative history of chaos theory (*Chaos: Making a New Science*, 1987), does not acknowledge that more than one branch exists. He barely mentions Prigogine's name in passing, describing his work as springing from a "highly individual, philosophical view" (p. 339). This remarkable omission testifies to how contested the name of chaos is, even within the physical sciences. (Some researchers in dynamical systems theory think that Gleick went too far in calling it a new science.) Nevertheless, there are points of convergence between the two branches. For example, the Belousov-Zhabotinskii reaction, which serves as a prime example of a self-organizing system, has also been shown to contain a strange attractor. In the face of these commonalities, that so definite a breach should exist has interesting political as well as philosophical dimensions, some of which will be touched upon in chapter 4, where Prigogine's work is discussed.

Despite the breach, it is possible to identify several characteristics that chaotic systems share. These characteristics will be discussed in detail in later chapters; it may be useful to indicate briefly here what they are. Perhaps the most general is *nonlinearity*. With linear equations, the magnitudes of cause and effect generally correspond. Small causes give rise to small effects, large causes to large effects. Linearity connotes this kind of proportionality. Equations that demonstrate it can be mapped as straight lines or planes.

Nonlinear functions, by contrast, connote an often startling incongruity between cause and effect, so that a small cause can give rise to a large effect. There is a good reason why linear equations have dominated the study of dynamical systems: nonlinear differential equations do not generally have explicit solutions. If nonlinear equations are introduced at all into physics courses, they are frequently relegated to the final hectic week or two of the course. The practice reinforces the assumption, implicit in Newtonian mechanics and encoded within the linguistic structure of stem and prefix, that linearity is the rule of nature, nonlinearity the exception. Chaos theory has revealed that in fact the opposite is true. To illustrate the predominance of nonlinear systems, Gleick quotes the mathematician

Stanislaw Ulam's quip that calling the science of chaos the study of nonlinear systems is like calling zoology the "study of nonelephant animals" (p. 68).

Take the weather. Even with large computers, it is impossible to predict weather patterns accurately more than a couple of days in advance. Weather prediction is difficult because small fluctuations quickly amplify into large-scale changes. Gleick estimates that if sensing devices were placed at one-foot intervals on poles as high as the atmosphere, and these poles were placed one foot apart all over the earth and the results fed into a supercomputer, the weather could still not be accurately predicted. In between the sensing devices, fluctuations in temperature and wind velocity would go unrecorded, and these fluctuations could soon affect global weather patterns (Gleick, 1987:21). The ability of minute fluctuations to cause large-scale changes holds for a wide variety of systems, from cream swirling in coffee to the thundering turbulence of Niagara Falls. Chaos is all around us, even in the swinging pendulum that for the eighteenth century was emblematic of a clockwork universe.[2]

Another characteristic of chaotic systems is contributed by their *complex forms*, which lead to a new awareness of the importance of scale. The new models bring into question an assumption so deeply woven into classical paradigms that it is difficult to see that it *is* an assumption. In classical physics, objects are considered to be independent of the scale chosen to measure them. A circle is assumed to have a set circumference, whether it is measured with a yardstick or with a ruler an inch long. Classical paradigms grant that smaller rulers may yield more precision than larger ones. But these differences are considered to be merely empirical variations that do not affect the existence of a "true" answer. This assumption works well for regular forms, such as circles, rectangles, and triangles. It does not work well for complex irregular forms—coastlines, for example, or mountain landscapes, or the complex branchings of the human vascular system. Here measurements on scales of different lengths do not converge to a limit but continue to increase as measurement scales decrease. Fractal geometry (a neologism coined by Benoit Mandelbrot, its inventor) expresses this complexity through in-

[2]For a discussion of chaotic pendulum swings, see Tritton, 1986.

creased dimensionality. Fractal forms possess additional fractions of dimensions, with the fractionality corresponding to the degree of roughness or irregularity in the figure.

Nonlinear dynamics, another important area within the sciences of chaos, is akin to fractal geometry in that it posits a qualitative and not merely a quantitative difference between linear and complex systems. Turbulent flow, for example, possesses so many coupled degrees of freedom that even the new supercomputers are inadequate to handle the required calculations. Since doing more of the same kind of calculations that one would use for laminar flow does not usually yield a solution, the difference between turbulent and laminar flow amounts to a qualitative distinction, an indication that another kind of approach is needed.

An essential component of this approach is a shift in focus from the individual unit to *recursive symmetries between scale levels*. For example, turbulent flow can be modeled as small swirls within larger swirls, nested in turn within still larger swirls. Rather than trying to follow an individual molecule, as one might for laminar flows, this approach models turbulence through symmetries that are replicated over many scale levels. The different levels are considered to be connected through coupling points. At any one of these coupling points, minute fluctuations can cause the flow to evolve differently, so that it is impossible to predict how the system will behave. The authors of a recent article on chaos in *Scientific American* illustrate the point by estimating that if an effect as small as the gravitational pull of an electron at the edge of the galaxy is neglected, the trajectories of colliding billiard balls become unpredictable *within one minute*.[3]

The comparison of chaos theory with quantum mechanics is illuminating in this respect. As is well known, quantum mechanics has important implications for the precision with which subatomic reality can be known. However, these implications are largely irrelevant for macroscopic bodies that remain stationary or move at ordinary speeds. The uncertainty principle does not alter the fact that tables and chairs stay neatly in place unless they are moved. Chaos theory, by contrast, studies systems configured so as to bring small uncertainties quickly up to macroscopic expression. Even a microscopic

[3]Crutchfield, Farmer, Packard, and Shaw, 1986.

fluctuation can send a chaotic system off in a new direction. This observation leads to another important characteristic of complex systems—their *sensitivity to initial conditions*. Quantum mechanics dovetails with chaos theory because it guarantees that there will always be some minimal level of fluctuation. An appropriately configured system will amplify these initial uncertainties until they are evident even on the macroscale.

To see why some systems magnify uncertainties and others do not, compare pushing alphabet blocks around with trying to get a rapidly moving bowling ball to move smoothly down the lane. Small hand tremors do not affect one's ability to position the blocks precisely, for the blocks have relatively large stabilizing surfaces. The bowling ball, by contrast, touches the alley at only a small portion of its surface, and its curvature makes it extremely sensitive to small motion tremors. Bowling is a difficult sport because a ball thrown in nearly identical ways can nevertheless follow very different paths. One time it may curve just right for a strike and another time veer off into the gutter, even though it was thrown almost the same both times. This extreme sensitivity to initial conditions is characteristic of chaotic systems. In fact, unless the starting conditions can be specified with *infinite precision*, chaotic systems quickly become unpredictable. Chaotic systems thus combine qualities that classical science considered antithetical and that quantum mechanics does not anticipate. Chaotic systems are both deterministic and unpredictable.

Other characteristics that complex systems share are *feedback mechanisms* that create loops in which output feeds back into the system as input. In certain chemical reactions, for example, a product may also serve as a catalyst for the reaction, driving it to generate more product, which in turn becomes more catalyst. The resulting dynamics are instrumental in explaining why organized structures can spontaneously emerge from initially small perturbations in the solution. In computer modeling of mathematical functions, iteration operates according to a similar principle, the output of one calculation serving as input for the next. When the function is strongly nonlinear, small fluctuations in the data are not smoothed out as iteration proceeds. Rather, they are magnified through a cascading series of bifurcations. In physical systems similar mechanisms are associated with the onset of turbulence, so that microscopic fluctuations can be amplified until they affect macroscopic flow.

The orderly disorder of chaotic systems had no recognized place within classical mechanics. By demonstrating that such systems not only exist but are common, chaos theory has in effect opened up, or more precisely brought into view, a third territory that lies between order and disorder. Does this mean that classical ideas of order have been repudiated? Here I think we must be cautious in drawing inferences about what the new sciences imply for the humanities. Chaos theory has a double edge that makes appropriations of it problematic for humanistic arguments that want to oppose it to totalizing views. On the one hand, chaos theory implies that Newtonian mechanism is much more limited in its applicability than Laplace supposed. On the other hand, it aims to tame the unruliness of turbulence by bringing it within the scope of mathematical modeling and scientific theory. It promises to provide at least some of the answers that Sir Horace Lamb, quoted in the epigraph, doubted even God could provide. From this perspective, chaos theory does not undermine an omniscient view. Rather, it extends it beyond where even Newtonian mechanics could reach. In this respect it is profoundly unlike most poststructuralist literary theories, especially deconstruction.

I am consequently wary of the claim that chaos theory provides confirmation from within the physical sciences that totalizing perspectives are no longer valid.[4] In my view, this inference at once overestimates and underestimates the significance of chaos theory. It is an overestimation because chaos theory is not opposed to normal science; it *is* normal science. It follows the same procedure as any other scientific discipline in accrediting its members, devising research protocols, and evaluating results. Like other sciences, it agrees that parsimonious explanations are better than multiplicitous ones. Mitchell Feigenbaum's name for his discovery that nonlinear

[4]See, for example, the discussion of "paralogy" in Lyotard, 1984. Alvin Toffler comes close to saying the same thing in his Foreword to Prigogine and Stenger, 1984:xi–xxvi.

dissipative systems follow a predictable path to chaos indicates how far the sciences of chaos are from the critique of universalism within the humanities: he called it universality theory.[5]

Nevertheless, it would be a mistake to think that chaos theory has no significant consequences for the humanities. On a deep level, it embodies assumptions that bring into question presuppositions that have underlain scientific conceptualizations for the last three hundred years. Among the challenges it poses are whether an effect is proportional to a cause; how scale variance affects facts; what it means to model a physical system; and what cosmological scenario we are taking part in. Through its concern with the conditions that make movement from local sites to global systems possible, it exposes presuppositions within older paradigms that made universalization appear axiomatic. In this sense it is an underestimation to say that chaos theory has redefined what science means. Changed are not the disciplinary procedures and criteria of normal science but the epistemic ground on which it—and much else in contemporary culture—rests.

When a dichotomy as central to Western thought as order/disorder is destabilized, it is no exaggeration to say that a major fault line has developed in the episteme. It would be strange indeed if there were not other theoretical enterprises that also work this fault line.[6] Some of the most visible are within poststructuralism, especially deconstruction. Just as the new scientific paradigms challenge the primacy traditionally accorded to ordered systems, so deconstruction exposes the interrelation between traditional ideas of order and oppressive ideologies. The scientific theories show that deterministic physical systems become chaotic because initial conditions cannot be specified with infinite accuracy; deconstructive readings operate upon texts to reveal the indeterminacy that results from the lack of an absolute ground for language. The scientific paradigms embody a shift of perspective away from the individual unit to recursive sym-

[5]Feigenbaum, 1980. It is not a coincidence that Einstein also considered naming his theory "universal invariance" before settling on "relativity."

[6]Among the important challenges to totalizing views which I will not discuss are Richard Rorty's *Philosophy and the Mirror of Nature* (Princeton: Princeton University Press, 1979) and *Consequences of Pragmatism: Essays, 1972–1980* (Minneapolis: University of Minnesota Press, 1982).

metries; poststructuralism, especially as exemplified in the work of Michel Foucault, writes about the death of the subject and the couplings that allow changes to sweep rapidly through a culture. The science of chaos reveals a territory that cannot be assimilated into either order or disorder; deconstruction detects a trace that cannot be assimilated into the binary oppositions it deconstructs. These correspondences are not accidental. They reflect what Christine Froula (1985), in comparing deconstruction with quantum mechanics, identified as a deepening crisis of representation in Western thought.

Because I discussed the scientific models first, I may have given the impression that they stimulated the development of analogous theories in literature. This is almost certainly not the case. The literary theories appeared simultaneously with, or slightly before, cognate formations in science. Postmodernism was already well articulated within contemporary literature and critical theory by 1970 and has precursors at least as early as Nietzsche. By comparison, Lorenz's seminal paper "Deterministic Nonperiodic Flow," first published in 1963, did not attract widespread attention (as indicated by frequency of citation) until about a decade later.[7] The chronology does not correspond with the tacit acceptance of scientific priority which for many years informed influence studies in literature and science (an acceptance I have challenged elsewhere as a reinscription of the scientism prevalent throughout the academy in the 1930s and 1940s).[8] From my point of view, an additional attraction of the new paradigms is their built-in resistance to this traditional mode of explanation. Even if I did not favor a cultural explanation, I would be thrown back upon it by the facts of the case, unless I were willing to grant that the correspondences are coincidence.

I have been construing the division between literature and science as the difference between literary theory on the one hand and scientific models on the other. Now I want to cleave the copula in

[7] Gleick points out that Lorenz's paper was not noticed for some time (1987:31).

[8] N. Katherine Hayles, "Reconfiguring Literature and Science: From Supplementarity to Complementarity," in *First-Born Affinities: Theories of Literature and Theories of Science*, ed. Paul Privateer and George Rousseau (Carbondale: Southern Illinois University Press, forthcoming), and "Turbulence in Literature and Science: Questions of Influence," in *American Literature and Science*, ed. Robert Scholnick (Chapel Hill: University of North Carolina Press, forthcoming).

another way, putting critical and scientific theory on one side and literature on the other. Whereas the earlier division facilitated comparisons between literary and scientific concepts, this construction brings into play distinctions between disciplinary and nondisciplinary work. To work within a discipline is to be trained in such a way as to absorb the practices, knowledge, and presuppositions that define the discipline. Among the practices that maintain and replicate disciplinary presuppositions are graduate advisory systems, course contents and selections, comprehensive examinations, and dissertation defenses. Different as literature and science are, they are both clearly disciplines in this sense.

Creative writers also work within traditions and read one another's work. But because they are not accredited in the same way, they have considerably more latitude in defining the contexts appropriate to evaluation of their work. Even for writers who come out of university writers' workshops (such as the one at the University of Iowa, where I teach), there is a very significant difference between the command they are expected to demonstrate over disciplinary traditions and that required from a graduate student in the English department. The accreditation most important for the creative writer cannot, in fact, be conferred by an institution. It comes largely from presses and magazines willing to buy the work, and thus from readers rather than peers and colleagues. This orientation toward readership and the relative freedom from academic specialization make creative writing more responsive to the culture in general than a strictly defined disciplinary tradition. As a result, it tends to carry with it more of the substrata of earlier cultural formations than does contemporary writing within disciplines.

I risk belaboring this rather obvious point because it has important implications for the way literature, as distinct from disciplinary writing, interacts with the new paradigms. Responding to a more diverse audience, creative writers are forced into an awareness of how stratified and heterogeneous culture is in a way that disciplinary writers are not. A physicist who works in plasma physics told me that perhaps three hundred people in the country read and understood his work, and that he personally knew 90 percent of them. Many literary critics could say the same. But what creative writer, if she hopes to thrive professionally, could afford to appeal to an audience this small and specialized?

Moreover, creative writing is located within complex fields of intertextual resonances that affect signification not only in the narrow sense of the way words are understood but also in the broader sense of the way plots are structured, characters conceived, actions represented. Combined, these factors make the literary texts more concerned than either chaos theory or deconstruction with the aura of cultural meanings that surrounds chaos. It is therefore important for my discussion of literature (and also for literary theory and science) to explore what chaos meant in its traditional guise as mythopoeic concept, before it acquired its contemporary connotations of maximum information, dissipative reorganization, and deeply encoded structure. These older resonances modify and partially transfigure new versions of chaos in literature, even as they are in the process of being articulated.

Within the Western tradition, chaos was associated with the unformed, the unthought, the unfilled, the unordered. Hesiod in the *Theogony* designates Chaos as that which existed before anything else, when the universe was in a completely undifferentiated state. Later in the *Theogony*, he uses the term to signify the gap that appeared when Heaven separated from Earth through the influence of Eros. Chaos and Eros thus have a mysterious connection as the two primeval forces of the world, although Chaos is the older of the two. According to Hesiod, Heaven and Earth embraced, rain fell, and the wet and dry, cold and hot became distinct from one another, thereafter combining in various proportions to form the universe. Narrating the birth of the world as a story of increasing differentiation of form, the *Theogony* depicts chaos both as not-form and the background against which the creation of form takes place.

The tradition that identified Chaos as that which existed when the world did not continued at least through the Renaissance. In *Paradise Lost*, God creates the world not out of nothing but out of Chaos, the primeval *materia* of the universe before it was invested with spirit. Also continuing into the Renaissance was the apposition between Eros and Chaos as the two primeval forces of the world. It occurs in Shakespeare, for example, although the tonalities are significantly darker than in Hesiod. In Shakespeare a return to Chaos signals a failure of love, as though the universe were disintegrating back to where it was before Eros appeared. Othello tells Desdemona, "Perdition catch my soul, / But I do love thee! and when I love thee not, / Chaos is come again" (*Othello*, III.iii,90–92). Similarly, when Venus mourns Adonis in Shakespeare's poem, we are told that "beauty dead, black chaos comes again" (*Venus and Adonis*, l.1020). The catastrophic implications of a return to Chaos become explicit in Ulysses' great speech on degree, where "Chaos, when degree is suffocate, / Follows the choking" (*Troilus and Cressida*, I.iii.125–126). In the same vein Gloucester, soon to become Richard III, says that he is unsuited to be a "ladies man" because his misshapen body is "like to a chaos" (*Henry VI*, pt. 3, III.ii.161).

During the Renaissance there are also frequent references to chaos as a lack of differentiation, a gaping void, a confused mass, an "undigested lump" (as George Sandys calls it in his translation of Ovid's *Metamorphosis*).[9] Sir Thomas Elyot warns in *The Governor* that if we "take awaie Order frome all thinges, what shulde than remaine? Certes nothing finally, except some man wold imagine eftesoones, Chaos, whiche of some is expounded, a confuse mixture."[10] The *Prymer, or boke of Private Prayer* (1560) perhaps says it best: "That old confusion, which we call chaos, wherein without order, without fashion, confusedly lay the discordant seeds of things."[11]

After the Renaissance, the classical suggestion that Chaos is the most ancient of all gods, the companion of Eros, and the stuff from which the world was made gradually grew obscure, and it finally dropped out of sight during the late eighteenth century. Enough of the classical resonances cling to Chaos for Pope to posit it as a deity (although of stunningly undistinguished lineage) in *The Dunciad*, where Dullness is described as "Daughter of Chaos and eternal

[9]George Sandys, *Ovid's Metamorphosis, Englished, Mythologized, and Represented in Figures*, ed. Karl K. Hulley and Stanley Vandersall (Lincoln: University of Nebraska Press, 1970), bk. 1, p. 25.

[10]Sir Thomas Elyot, *The boke named The gouvenour*, ed. Henry Herbert Stephan Croft (London: C. K. Paul, 1880), p. 3.

[11]Cited in the *Oxford English Dictionary*, ed. C. T. Onions et al. (Oxford: Clarendon, 1933), vol. 3, under "Chaos." The date given by the *Short-Title Catalogue* for this edition is 1560, not 1559, as the *OED* has it.

Night" (I.12). But the shift in tone is significant, and few references to Chaos as a god persist beyond that.

Slowly taking the place of this view of chaos was one that envisioned it as the antagonist to order. By the early nineteenth century, this view was predominant. Consider Edgar Allan Poe's tale "The Fall of the House of Usher" (1839). The preternatural and incestuously erotic energy that animates Madeline as she struggles to emerge from her coffin is in one sense an organizing force, for it brings her back to a living state; but these very exertions ensure that her approaching dissolution will be final, precipitating Roderick's death as well as hers. Madeline's exertions have their analogue in the storm that seems to animate the miasmic fog around the house. Just as Madeline's energy leads to the death of her twin, so the storm's expenditure of energy is linked with the disintegration of the house and the end of the family it sheltered and signified. These correlations imply that order and chaos are bound together in a dialectic. The more energy expended, the more certain the collapse into fragmentation and chaos. In the isolated and incestuously closed system formed by the House of Usher, Poe anticipated the second law of thermodynamics, which was being formulated by Sadi Carnot and others about this time.[12]

The popularization of thermodynamics during the 1860s and 1870s reinforced the antagonistic connection between order and chaos through predictions of a cosmic dissipation that would end with all heat sources everywhere being exhausted, resulting in the so-called "heat death" of the universe. Countering this pessimistic scenario was the awareness that in the short run (that is, in the eons while life still continued on earth), the release of thermal energy could run trains, fuel steamships, generate electricity. The power of heat engines was of a different kind as well as a different order from the mechanical advantage bestowed by pulleys, levers, and winches.

[12]Sadi Carnot came close to formulating the second law in his 1824 treatise, *Réflexions sur la puissance motrice du feu et sur les machines propres à développer cette puissance* [Reflections on the motive power of fire and on the proper machines for developing this power]. In 1849 Lord Kelvin (William Thomson) wrote a paper that jointly attributed the second law to Carnot and to Rudolf Clausius, who pointed out the error that kept Carnot from arriving at a fully consistent formulation of the second law.

Whereas mechanical energy was orderly, conveyed along visible lines of force, thermal energy was chaotic, driven by the fiery turbulence of the furnace and deriving its power from the random motions of molecules whizzing at unbelievable speeds. In "Turner Translates Carnot," Michel Serres argues that this vision of a turbulent, chaotic, immensely powerful energy inspired many of Turner's paintings (1982a, 54–64).

The ambiguity that once inhered in the concept of chaos was thus reconceptualized during the nineteenth century as a tension between a short-term release of energy and a long-term price paid for that release. Popularizations of this formula often took considerable liberties with interpretations of "long-term" and "short-term". Take Henry Adams. In *The Education of Henry Adams*, substantially completed in 1907, Adams intimates that as man captures more and more powerful energies, history accelerates accordingly, until human intelligence (and implicitly the world) flies apart because it can no longer stand the strain. Yet in "A Letter to American Teachers of History," completed only three years later, he argues that human intellect has become progressively degraded since the Middle Ages, and that this degradation is inevitable because humans tap more and more powerful energies to keep them going. We see that the formula could be appropriated and used in very different ways, even by the same author.

Starting after World War I, and increasingly after World War II, the energy/dissipation ambiguity within chaos was shadowed by a corresponding ambiguity within order. On the one hand, order connoted stability, regularity, predictability. On the other, it signified a directive or a symbolic configuration one is not free to disobey, as in a military order or Foucault's "order of things" (1970). As chaos came to be seen as a liberating force, order became correspondingly inimical, associated with the mindless replication of military logic or with the oppressive control of a totalitarian state (or state of mind).

By 1960, the stage was set for chaos to undergo a radical reevaluation. However, the multiple meanings it had acquired through centuries of commentary made it capable of diverse significations, even when it was cast in a generally positive light. This polysemy is evident in the differences between literary and scientific valuations of chaos. Literary theorists value chaos primarily because they are preoccupied with exposing the ideological underpinnings of traditional ideas of order. They like chaos because they see it as opposed to order. Chaos theorists, by contrast, value chaos as the engine that drives a system toward a more complex kind of order. They like chaos because it makes order possible.

The examples illustrate how the cultural field can energize different kinds of reevaluations, because the arrows in the field do not all point in the same direction. They also suggest that there is an interaction between the culture and the new paradigms, rather than a one-way flow of influence from one to the other. Even as the new paradigms bring into focus classical texts that may not have fitted very well into older traditions, these texts help to give traditional authority to the new paradigms. The idea that chaos could give rise to order, for example, is not without precedent. Henry Adams played with it; Poe ambiguously inscribed it; Hesiod gave it mythic authority. If these resonances had not existed, I think it unlikely that chaos theory would have aroused the widespread interest that it has. The entirely new concept is more likely to pass unheralded and unrecognized than one whose avatars echo through time.[13]

I suggested earlier that creative writing tends to carry along more of the cultural substrata than does writing within disciplines. Many postmodern texts self-reflexively play with their conditions of possibility, flaunting the stratification that in any case they are unlikely to escape. Consider Italo Calvino's *Cosmicomics* (1968). The stories in this collection are narrated by Qfwfq, who speaks with a human voice and consciousness and yet was present when the universe was not. The implication is that Qfwfq has evolved along with the universe, eventually becoming a being like us, and so able to communicate in a language we can understand. But the language that he (it?) uses is radically at odds with his attempt to tell us what it was like, for language is a relatively recent invention in the universe's history. Hence Qfwfq is constantly forced to qualify his descriptions, using words to describe what existed before language itself was born. Within this wildly anachronistic project, traces of ancient cos-

[13]This point is made in several of the essays in Borges, 1964, including "Pascal's Sphere," "Kafka and His Precursors," "Avatars of the Tortoise," and "The Enigma of Edward FitzGerald."

mogonic myths coexist with ideas taken from the new scientific paradigms.

In "The Form of Space," the narrator (presumably Qfwfq, although he is not identified) falls in a straight line through space, parallel to the fall of Ursula H'x, whom he desires, and Lieutenant Fenimore, whom he fears and detests as a potential rival. Throughout his fall, so reminiscent of the movement of Democritus's atoms, the narrator dreams of a swerve or clinamen that would bring him into contact with Ursula. Following the reasoning of Lucretius in *De Rerum Natura*, the narrator thinks of the clinamen as a longed-for release from the deadly order of the same.[14]

Significantly, he finds release not in actuality but in the curved lines of the letters constituting the text that writes him. He understands that if the lines of this writing were parallel, like his fall and those of his companions, signification would be impossible and hence nothing would happen in the story. Indeed, there could be no story for anything to happen in—no Ursula, no lieutenant, no narrator. The curvature of writing is thus equated with the swerve that Serres identifies with chaotic turbulence. Writing is turbulence, or more precisely, brings turbulence into being. The association suggests that before the world could exist, there had to be chaos; and simultaneously with chaos came writing, in the swerving inscriptions of the atoms when they first deviated from their linear paths. From this point of view, anachronism is not so much a fallacy as an inevitability. The story illustrates how contemporary ideas of chaos and signification can merge with ancient beliefs to form a narrative almost as stratified as culture itself.

It is this kind of play between new ideas and traditional formations that we are in danger of missing if we think of the present epistemic shift as homogeneous or total. The new paradigms break with old presuppositions; but language, tradition, and culture also affect our understanding of what is new in them. Working sometimes with, sometimes against the break with the past are the dynamics peculiar to a given site—the recursive techniques of non-

[14]Compare "Lucretius: Science and Religion," in Serres, 1982a:98–124. Whether Calvino's inspiration came directly from Lucretius or was mediated through Serres, I do not know.

linear dynamics, the narrative conventions that a novelist such as Doris Lessing inherited from such predecessors as Conrad, James, Woolf, and Joyce, the anxiety of influence among critics when Derrida appeared on the scene. The resulting flow is turbulent rather than laminar, circling around in eddies and backwaters, responding to disturbances that affect the macroscopic qualities of the stream as it splits into tributary streams or converges into a swelling flood. For this reason my own narrative is nonlinear as it moves back and forth among scientific models, critical theory, and literary texts, following now one stream and now another, but always edging toward the alluvial plain where the currents run together to create the phenomena we call postmodernisms.

Subsequent chapters, reflecting the division between the two branches of chaos theory, are divided between texts that see chaos as a void from which something can emerge and those that see chaos as a complex configuration within which order is implicitly encoded. The first grouping, headed "Something out of Nothing," begins with a chapter that traces the transvaluation of entropy as it moves from classical thermodynamics to information theory. It shows that a crucial step in the reevaluation of chaos was the interpretation of randomness as maximum information. Chapter 3, on *The Education of Henry Adams*, shows similar ideas at work in a literary text. In *The Education*, the chaotic void becomes the space of creation. Through a combination of historical theorizing, personal trauma, and narrative strategies, *The Education* transforms its voids and ruptures into gaps from which radiates an energy that radically reorganizes whatever comes in contact with it. This reorganization is by no means unequivocally good; but it is finally irresistible, and therefore irrevocably bound up with human history.

Chapter 4 discusses the work of Prigogine and Stengers, focusing on their claim that the new paradigm reconciles being with becoming. This chapter, in addition to laying out the scientific basis for the order-out-of-chaos concept in irreversible thermodynamics and cosmology, speculates about the role of vision in shaping and directing scientific research programs. The "Something out of Nothing" section concludes with Chapter 5, on Stanislaw Lem. Like Adams, Lem conceives of chaos as a void from which something emerges. However, he works out a unique explanation for the creative power of chaos, seeing it as bound together with order in a complex dialectic through which chaos and order come to interpenetrate each other without losing their distinctive identities. In this respect Lem is poised on the threshold between the old and new paradigms, not going so far as to redefine chaos but believing that it must somehow be incorporated into our picture of the world. His work illustrates that transitions to new paradigms are never homogeneous or uniquely determined. Even when many of the same elements are present, they can be combined differently by writers with different backgrounds and agendas.

The second group of chapters, headed "The Figure in the Carpet," begins with a discussion of the strange-attractor branch of chaos theory. Among the theories discussed in chapter 6 are Feigenbaum's universality theory, Mandelbrot's fractal geometry, Robert Shaw's interpretations of chaos as information, and Kenneth Wilson's quantum field theory. The chapter concludes with a discussion of James Gleick's *Chaos*, showing how its narrative patterns conflict with some of the new paradigm's fundamental assumptions. Chapter 7 discusses poststructuralist theories, focusing on the work of Jacques Derrida, Roland Barthes, and Michel Serres. It experiments with creating a model that can account for divergences as well as isomorphisms between fields. It suggests that analogies are best understood through an ecology of ideas, and that differences can be accounted for through various disciplinary economies and the equivocations that result when a concept is imported from one discipline to another.

Chapter 8 addresses the politics of chaos. Among the ideological issues raised by the new paradigms are the valorization of local knowledge and the corresponding repudiation of global theories. But the situation is more complex than it may at first appear. Chaos theory, although it recognizes the importance of scale, does so in order to globalize more effectively. Similarly, poststructuralism finds a new globalizing imperative in its insistence that there can be no global theories. These complexities suggest that ideological stances cannot be fully understood apart from the disciplinary contexts in which they are embedded.

Chapter 9 deals with Doris Lessing's *Golden Notebook* (1962). The politics of chaos are central to the text's narrative organization, as well as to its wide-ranging analyses of Marxism, sexism, and racism. *The Golden Notebook* has many of the characteristics associated with the new paradigms—a problematic relation between local sites and global theories, an interest in recursive symmetries as a principle of organization, an awareness of how small fluctuations can effect large-scale changes. These analogies cannot be explained by scientific influence, for Lessing knew little or nothing of chaos theory. Rather, they represent the independent re-creation by Lessing of a chaos theory of her own. Convinced that society is on the verge of cracking up, Lessing's narrator desperately tries to find clues to a new synthesis in her own disintegrating psyche. This text thus provides insight into the broad cultural movements that underlie the new paradigms.

The two major sections, "Something out of Nothing" and "The Figure in the Carpet," are framed by the present introductory chapter and a concluding chapter that locates chaos theory within the narrative of postmodernism. Chapter 10 argues that postmodernism can be understood as a continuing process of denaturing, that is, of realizing that concepts once considered natural are in fact social constructions. The denatured concepts include language, time, context, and, increasingly as postmodernism progresses, the human. Chaos has its frightening as well as its liberating aspects. Fragmentation and unpredictability are not, I argue, always cause for celebration.

The title of this study, *Chaos Bound*, was chosen because it hints at the complexities of these contemporary inscriptions of chaos. One cluster of meanings centers on the sense of "bound" as a limit line or boundary. Bringing into view a third territory that lies between order and disorder, chaos theory draws boundaries where previously there was only bifurcation. It also emphasizes the importance of boundaries within systems.

A second set of meanings comes into play with the sense of "bound" as confinement or bondage. In *The Postmodern Condition* (1984) Lyotard sees some aspects of chaos models, particularly fractal geometry, as promising a release from totalizing narratives. Yet other aspects of chaos theory bring this vision into question, for they embody a shift in perspective away from the individual and toward systemic organization. As we move into an episteme that sees the human as a social construct, it is an open question whether individuality, in its traditional Western guises, can continue to be seen as a liberating or energizing force.

A third cluster of meanings emerges from the sense of "bound" as resolve and determination, especially to attain a destination. The emergence of the new scientific paradigms are authorized by the same cultural developments that have led to literary and critical postmodernisms. At once promising an escape from oppressive order and implying that even chaos can be orderly, these theories replicate the ambiguity at the heart of postmodern culture. Whether arrival at this destination will transform society, as it has already transformed some scientific and literary sites, is a question beyond the bounds of *Chaos Bound*. Nevertheless, it seems clear that local fluctuations are in the process of being magnified into large-scale changes. Chaos both exemplifies this process and names it.

Part I

SOMETHING OUT OF NOTHING

CHAPTER 2

Self-reflexive Metaphors in Maxwell's Demon and Shannon's Choice: Finding the Passages

In a sense, all language is metaphoric. When a carpenter says that a room is 7 yards long, he is comparing the length of the room with the length of an Anglo-Saxon girdle. When a scientist says that a molecule has a diameter of 2.5 angstroms, the standard has changed but the principle is the same; the object is still understood in terms of its relation to something else.[1] A completely unique object, if such a thing were imaginable, could not be described. Lacking metaphoric connections, it would remain inexpressible.[2] The question is thus not whether metaphors are used in science as well as literature, but rather how metaphors are constituted in the two disciplines, how they change through time, and how they are affected by the interpretive traditions in which they are embedded.

[1]This point is made by George Lakoff and Mark Johnson in *Metaphors We Live By* (Chicago: University Of Chicago Press, 1981). They emphasize that the "essence of metaphor *is understanding and experiencing one kind of thing in terms of another* (p. 5). I am also indebted to F. C. McGrath's manuscript "How Metaphor Works: What Boyle's Law and Shakespeare's 73rd Sonnet Have in Common." Also important in informing my remarks on metaphor is Max Black's definitive study, *Models and Metaphors: Studies in Language and Philosophy* (Ithaca: Cornell University Press, 1962).

[2]Gillian Beer's fine discussion of metaphor in Darwin and his contemporaries also makes this observation: *Darwin's Plots: Evolutionary Narrative in Darwin, George Eliot, and Nineteenth-Century Fiction* (London: Routledge & Kegan Paul, 1983). I am indebted throughout this essay to Beer's insights.

In discussing how metaphors work, Paul Ricoeur points out in *Interpretation Theory: Discourse and the Surplus of Meaning* (1976) that it is misleading to analyze a metaphor at the level of individual words, for the essence of a metaphor is the relation it establishes between words (pp. 46–52). A metaphor posits a connection rather than a congruence. It points to a similarity, but the similarity is striking because in other respects the concepts are very different. A metaphor is vital only as long as the relation is problematic—that is to say, as long as similarity and difference are both perceived to be present. When differences in the relation have been so successfully suppressed through use and habit that they are no longer capable of putting a torque or, as Ricoeur says, a "twist" (p. 50) on our understanding of the concepts, the metaphor is dead. "There are no live metaphors in a dictionary," Ricoeur asserts (p. 52). He is not quite correct, for metaphors that appear to be dead may be brought back to active tension again through their interplay with the surrounding context, as the split writing of deconstruction has taught us. Dead metaphors too easily suggest corpses that can be safely buried and forgotten. Rather than thinking of metaphors as dead or alive (adjectives that are themselves dead as metaphors), I prefer to consider them dormant or active. The distinction is important to the story I have to tell, for it is a narrative of metaphors expanding and collapsing, fading into dormancy and being tightened into tension by changing cultural contexts in interplay with disciplinary traditions.

The story begins with a thought experiment proposed by James Clerk Maxwell in 1879 which came to be known as "Maxwell's Demon." Maxwell's Demon is one of the most famous conundrums in the history of science, provoking over a hundred years of commentary, interpretation, revision, and speculation. It sparked a crucial development in information theory, marking the inscription into scientific discourse of a new attitude toward chaos and disorder. This development I call "Shannon's Choice," after the decision that Claude Shannon made to equate information with entropy, rather than to oppose them, as had been accepted practice until then. This decision was possible, I argue, because the underlying heuristic had changed, allowing the problem that Maxwell's Demon had posed to be conceptualized in a new way. The example illustrates how scientific theory can be guided by conceptual sets embodied in heuristic narratives, even though these sets are not part of the theory as such. If this is true, then metaphor may play a larger and more active role in scientific theorizing than has hitherto been recognized.

Heuristic fictions such as Maxwell's Demon are like metaphors in that they posit a relation between the fiction and the theory which gestures toward similarity at the same time that it encounters the resistance of difference. The loose bagginess of the fit between the heuristic fiction and the theory is important, for it can open passages to new interpretations. Equally important is the language used to construct the heuristic. Following Max Black, Ricoeur observes that "to describe a domain of reality in terms of an imaginary theoretical model is a way of seeing things differently by changing our language about the subject of our investigation" (p. 67). The detour through language which the heuristic represents creates polysemous connections not present in the theory itself.[3] Overlaying a heuristic onto a theory is never merely an inert transposition of concepts, for it generates a surplus of signification that can lead to interpretations not intended by the person who proposed the theory or, for that matter, by the heuristic.

Within the redescription process that the heuristic entails, moments of special complexity appear when it contains a metaphor that self-reflexively mirrors the heuristic itself. A traditional visualization of a metaphor images it as a geometric compass, with one leg grounded and one leg moving free. The grounded leg alludes to the similarities between concepts brought into relation by the metaphor; the freely moving leg evokes the differences between them which can cause our understanding to land at unexpected points. A metaphor that self-reflexively mirrors itself in another metaphor threatens to lose the grounding that reassures us the comparison is not entirely free-floating. It could be imaged as a compass with one leg moving freely and the other resting not on ground but on the leg of another compass. Postmodern writers have exploited this lack of ground to reveal the intrinsic reflexivity of all language. Borges's emblem of a staircase that ends in space, leading not to a door but to vertigo,

[3]Mary Hesse draws attention to the importance of what she calls "redescription" in *Models and Analogies in Science* (Notre Dame: University of Notre Dame Press, 1966).

speaks to the dangerous potential of metaphors to expose the ungrounded nature of discourse.

From a scientist's viewpoint, the vertiginous staircase or balancing compass explains very well why metaphors have not been admitted as valid components of the scientific process. Already suspicious of the looseness that a freely moving compass leg would imply, they are even warier of a mode of speaking, thinking, and writing that can lose its ground entirely. What this response misses is the fact that language is always already metaphoric. If we do not feel vertigo, it is because long usage has inured us to balancing over the abyss. At moments of dangerous reflexivity, when the polysemy of metaphor threatens to overwhelm scientific denotation with too much ambiguity, the tradition confronts the new possibilities that metaphor has brought into play. At this point a bifurcation is likely to appear, for the situation is sufficiently complex so that even a small fluctuation can send the commentary surrounding the heuristic in a new direction. It is no accident that decisive turns in the traditions I discuss are often associated with self-reflexive metaphors.

In an early essay, "From Science to Literature" (1967), Roland Barthes distinguishes between science and literature through their different attitudes toward language. Science, Barthes says, regards language instrumentally. For science, language (which is nothing) serves only to transmit concepts (which are everything). In literature, language is not a vehicle transmitting the object, but the object itself. Barthes is interested in what happens to this dichotomy between literature and science when structuralism is injected into it. Structuralism prides itself on being a science but has its roots in linguistics. Derived "from linguistics, structuralism encounters in literature an object which is itself derived from language" (1986:6). The question Barthes poses is whether structuralism will (like a science) pose itself above its object or recognize that it is itself composed of the language it would take for its object. Anticipating the advent of deconstruction and other poststructuralist theories, Barthes predicts that structuralism "will never be anything but one more 'science'. . . if it cannot make its central enterprise the very subversion of scientific language. . . . [It must work to] abolish the distinction, born of logic, which makes the work into a language-object and science into a meta-language, and thereby to risk the illusory privilege attached by science to the ownership of a slave language" (p. 7).

Since Barthes wrote these lines, the project he outlined for structuralism has of course spread far beyond its boundaries.[4] So pervasive is the recognition today that language is never transparent that it seems almost quaint to associate the insight exclusively with structuralism. If structuralism has been superseded, however, the project Barthes set forth has not. The task of understanding how scientific languages are implicated in the concepts they convey remains one of the important problems of literature and science. To this project the study of self-reflexive metaphors can offer distinctive contributions, for at these moments science necessarily confronts the enfolding of language-as-object into its assumed stance as a metalanguage. That is to say, at these moments science confronts its literariness.

The self-reflexive metaphor is analogous to the encoding scheme that Kurt Gödel worked out in proving his incompleteness theorem, whereby statements about numbers were made simultaneously to function as statements about statements about numbers (Gödel, 1962). Gödel's theorem proved that it is always possible to devise a coding scheme (for formal systems complicated enough to allow arithmetic to be done in them) that will collapse the distance between statements and metastatements. The result of this conflation of a metalanguage with an object language is an inherent undecidability that defeats all attempts at formal closure. If we take Gödel's theorem as itself a metaphor for self-reflexive metaphors, it suggests that at these moments, an inherent undecidability emerges which cannot be resolved within the system itself. This is why reflexive metaphors often function as crossroads or junctures, for the undecidability opens up passages that lead in new directions.

[4]The question whether a metalanguage could be created which would not be contaminated by the assumptions of an object language was central to the unraveling of the positivist program, as well as to the attempted formalization of mathematics. In *The Structure of Scientific Inference* (Berkeley: University of California Press, 1974), Hesse addresses the question whether there is an independent observation language in science (pp. 9–45). Frederick Suppe's collection remains a landmark in the debate: *The Structure of Scientific Theories*, 2d ed. (Urbana: University of Illinois Press, 1977).

At this point I am obliged to address a difficulty in which I immediately become involved if I suggest that the above remarks constitute a theory that the rest of the chapter will apply in practice. Insofar as my comments about self-reflexive metaphors imply a theory, it is a theory about the impossibility of separating theory from practice, formal results from heuristics, language-as-concept from language-as-vehicle. To illustrate how my own practice already determines my theory even as I articulate it, consider the lines where I attempt to explain how metaphor works: they are shot through with metaphors, from corpses to compasses. Since I can explain metaphor only through metaphors, practice interpenetrates theory from the beginning—a situation I will henceforth recognize by putting "theory" in quotation marks. If my "theory" is correct, it implies that the attempt to arrive at a theory about literature and science is hopeless from the outset, for theory is always already determined by disciplinary practices that are necessarily different for literature than for science.[5]

In science, "theory" generally means a set of interrelated propositions that have predictive power and therefore have the potential to be refuted. In literature, by contrast, "theory" means a set of speculative statements that serve as guides to reading and interpreting texts. Literary critics do not attach much importance to the predictive power of literary theories, for most would agree that one's theoretical orientation determines what will be seen, at least in part. A literary "theory" falls into disuse not because it has been refuted but because its assumptions have become so visible to its practitioners that it can no longer effectively create the illusion that it is revealing something about the text that is intrinsically present, independent of its assumptions.

What does it mean, then, to posit a theory about literature and science? To answer this question, one would have to presuppose a set of disciplinary practices that constitute Literature and Science as

[5]Stanley Fish made a similar point in his paper on interdisciplinarity read at the 1988 Modern Language Association Convention, "Being Interdisciplinary Is So Very Hard to Do." According to Fish, the very idea of interdisciplinary work engenders an unresolvable paradox. How his own work in literature and law is possible he did not address, rather like the aeronautical engineer who proved that a bumblebee could not fly.

a field of its own. Supposing that such distinctive practices exist (a proposition I regard as problematic), the resulting "theory" will be different from theories about literature and theories about science. It will not, however, be a metatheory capable of subsuming theories in other disciplines. The only hope for a truly interdisciplinary theory, it seems to me, is a "theory" about the impossibility of creating a theory that will not be implicated in disciplinary practice. Such a "theory" is interdisciplinary not because it transcends disciplines but because it recognizes the rootedness of every theory in the discursive practices characteristic of its discipline.

I want to explore the implications of a "theory" about scientific metaphor by engaging in a recursive analysis that traces the play of surplus meaning in Maxwell's Demon until the interpretive tradition arrives at a self-reflexive moment, that is, a point where the heuristic becomes a metaphor for itself. At that moment is born a new interpretive fiction that differs fundamentally from the heuristic that fathered it. This fiction is Shannon's Choice, in which Shannon associates information with disorder rather than order. Shannon's Choice in turn engendered a self-reflexive metaphor that opened new passages, this time into the paradigm known as chaos theory. I use a narrative method because it is only when enough layers accrete that the self-reflexive metaphors can be revealed. Many readers may find this chapter "noisy" in its density of detail and close analysis of scientific concepts. The analysis is necessary, however, if the heuristics are to be rigorously understood. Einstein's remark about physics holds true for the history of science: God is in the details.

In retrospect, it appears that the self-reflexive moments I discuss acted like switches on a railroad track, sending the train of thought in a different direction. Or perhaps they occurred because the views they implied had been highly energized by cultural events, and the self-reflexive moments acted as conduits or fissures that allowed these ideas to erupt into the scientific traditions. However the self reflexive moments came about, they created instabilities within the heuristics which were amplified until they became vortices of turbulent signification. Out of these vortices, like Venus from the sea, arose new attitudes toward chaos which called into question the traditional relation of order to disorder, information to meaning, human understanding to that which it understands.

To appreciate the complexities of Maxwell's Demon, it is first necessary to understand entropy—a formidable task, because the word has undergone so many changes in meaning that it actually encompasses several concepts. It makes no sense to ask what entropy "is," as though it were possible to find in this protean signifier a transcendent signification. Instead we must ask what it meant to whom, for what reasons, in what context, and with what consequences.[6]

The word "entropy" entered the language when Rudolf Clausius coined it from a Greek word meaning transformation. To Clausius entropy was linked to the inevitable degradation of heat which occurs in any heat exchange (Clausius, 1850; Cardwell, 1971). Entropy is a measure of the heat lost for useful purposes. Suppose you boil water in a kettle, pour the water into a mug, and put your hands around the cup. Some of the stove's heat is lost to the air and kettle, some to the mug and your hands. When the heat exchanges in the room as a whole are taken into consideration, the total amount of heat remains nearly constant (if we neglect that lost to the environment outside the room). But after the transactions are complete, much of the heat is in such diffuse form that it can no longer be harnessed for useful purposes. Clausius expressed the fact that no energy is lost through the first law of thermodynamics, which states that the total amount of energy in a closed system remains constant. The first law says nothing, however, about the *form* in which this energy exists. This issue is addressed by the second law, which decrees that in a closed system entropy always tends to increase. The second law implies that no real heat transfer can be 100 percent efficient.[7]

It did not take scientists long to realize that if some energy is

[6]The changing meanings of "entropy" in thermodynamics, statistical mechanics, and information theory are reviewed in K. G. Denbigh and J. S. Denbigh, *Entropy in Relation to Incomplete Knowledge* (Cambridge: Cambridge University Press, 1985).

[7]Entropy is defined as heat divided by the absolute temperature, on a temperature scale whose zero point represents the lowest possible degree of heat. The caveat that no *real* heat transfer is 100 percent efficient is necessary because it is possible to imagine an idealized system in which changes occur with infinite slowness. When a system is always at equilibrium, the entropy remains constant. However, this condition cannot be attained in the real world. In all real heat transfers in a closed system, entropy increases.

always lost for useful purposes in every heat exchange, there will eventually come a time when no heat reservoir exists anywhere in the universe. At this point the universe experiences "heat death," a final state of equilibrium in which the temperature stabilizes near absolute zero (about −273° C) and there is no longer any heat differential to do work or sustain life. These implications of the second law were made explicit in 1852 by William Thomson (Lord Kelvin), the great British thermodynamicist. Kelvin (1881) summarized them in three "general conclusions." First, "there is at present in the material world a universal tendency to the dissipation of mechanical energy"; second, "any restoration of mechanical energy, without more than an equivalent of dissipation, is impossible . . . and is probably never effected by means of organized matter, either endowed with vegetable life or subjected to the will of an animated creature"; and third, "within a finite period of time . . . the earth must again be unfit for the habitation of man as at present constituted" (p. 514).

Implicit in Kelvin's rhetoric are connotations that link these scientific predictions with the complex connections among repressive morality, capital formation, and industrialization in Victorian society.[8] The convergence of social formation with scientific concept is registered through language, illustrating how language-as-instrument is always already enfolded into language-as-concept. The "universal tendency toward dissipation" places entropic heat loss in the same semantic category as deplorable personal habits. The reversal of this tendency requires a "restoration." But any attempt at reform only creates more dissipation. A *net* restoration is beyond the power of "organized matter"; the adjective implicitly acknowledges that matter may be unorganized, itself subject to entropic decay. As the passage builds to its climax, it moves up the chain of being. "Vegetable life" cannot conquer entropy; neither can "animate creatures," even though they possess wills capable of "subjecting" matter. As a result of these failures, "the earth *must again* be unfit for the habitation of man." Human existence is thus bracketed between a prehistoric past when the earth was too hot for habitation and a thermodynamic

[8]A persuasive argument for the cultural subtext of Kelvin's theory of heat is presented by Crosbie Smith, "Natural Philosophy and Thermodynamics: William Thomson and the 'Dynamical Theory of Heat,'" *British Journal of the Philosophy of Science*, 1 (1976): 293–319.

future when it will be too cold. As the earth proceeds along this irreversiblc path, man must inevitably perish if he remains "as at present constituted." If man is to escape this dismal prediction, some unimagined transformation will have to take place.

These connotations are embedded within a text and a discipline concerned with the transfer and conservation of heat—concerns that had direct application to the expansion of the British Empire; this is why thermodynamics is sometimes called the science of imperialism. To Kelvin and his fellow thermodynamicists, entropy represented the tendency of the universe to run down, despite the best efforts of British rectitude to prevent it from doing so. In Kelvin's prose, the rhetoric of imperialism confronts the inevitability of failure. In this context entropy represents an apparently inescapable limit on the human will to control.

The very slight margin of escape Kelvin allowed himself in his prediction was well advised, for the second law quickly became one of the most controversial results of thermodynamics. An important development came in 1859, when James Clerk Maxwell published a paper deriving the properties of a gas from the statistical spread of molecular velocities within the gas (Maxwell, 1860, 1890). The implications of Maxwell's methodology were not at first apparent, for it was thought that his statistical treatment was merely a convenient way of treating systems about which one has incomplete information. Later it was recognized as a philosophical landmark, because it supported the view that thermodynamic laws are statistical generalizations rather than laws in an absolute sense. According to this interpretation of the second law, there is nothing to prevent the air molecules in a room from clustering in one corner. Such an event has an infinitesimally low probability; calculations show that it is quite unlikely to happen once during the time the universe has been in existence. Small as it is, this tiny margin of improbability keeps the second law from having the force of absolute truth. Strict determinism thus yielded to probabilistic prediction in Maxwell's interpretation of entropy.

Another important step came when Ludwig Boltzmann (1909) extended Maxwell's statistical method to arrive at a more general understanding of entropy as a measure of the randomness or disorder in a closed system. Boltzmann calculated the entropy S as $S = k$(log

W), where *k* is a universal constant and *W* is the number of ways the system can be arranged to yield a specified state. Suppose we flip a coin four times, and for each set of four tosses record the results. Only one arrangement gives four heads—*HHHH*. But there are six ways to get two heads and two tails:

HHTT *HTHT* *HTTH*
TTHH *THTH* *THHT*

The quantity *W* in Boltzmann's formula is therefore larger for the two heads/two tails state than for four heads, because there are more ways to arrive at the mixed state. This result corresponds with our common-sense intuition that it is safer to bet that the cards in a poker hand are of different suits than that they are all of the same suit. The more mixed up or randomized the final state, the more probable it is, because the more configurations there are that lead to it. Thus in Boltzmann's formula the entropy increases with the probability of a given distribution, with the most dispersed being the most probable.

Although Boltzmann and Clausius interpreted entropy differently, it is possible to reconcile the two interpretations. Heat is essentially a measure of internal energy. For gases this energy correlates with the average speed of the molecules.[9] The hotter the gas, the more heat it loses as it undergoes heat exchange, and so the more entropic it is in Clausius's terms. It is also more entropic in Boltzmann's sense because the greater the average speed of the molecules is, the more mixed up they become. The two formulations are thus equivalent—but they are not identical. The statistical interpretation contains important implications that the heat formulation lacks. To think of entropy as a statistical measure of disorder allows its extension to systems that have nothing to do with heat engines. In fact, so rich in significance is the statistical view of entropy that its full implications

[9]Stephen Brush, in a private communication, pointed out that there will always be a spread of molecular velocities in a gas at any temperature above absolute zero. Thus even if gas *A* had a lower temperature than gas *B*, some molecules in *A* would be moving faster than the average speed of the molecules in *B*. By selecting only those molecules, the Demon could thus make heat flow from cold to hot, which is an even stronger violation of the second law than Maxwell suggested in *Theory of Heat*.

are still being explored. Its immediate consequence was to weaken further the absoluteness of a predicted "heat death" by giving entropy an interpretation that was overtly probabilistic rather than deterministic.

With this background, we are now ready to consider the thought experiment that Maxwell proposed to test the second law. In a short note near the end of *Theory of Heat* (1871), Maxwell envisioned a microscopic being who could separate fast molecules from slow ones in a closed system, and so decrease the system's entropy without doing work. So concise is Maxwell's description that it may be quoted directly.

> If we conceive a being whose faculties are so sharpened that he can follow every molecule in its course, such a being, whose attributes are still as essentially finite as our own, would be able to do what is impossible for us. . . . Now let us imagine [that a vessel full of air] is divided into two portions, *A* and *B*, by a division in which there is a small hole, and that a being, who can see the individual molecules, opens and closes the hole, so as to allow only the swifter molecules to pass from *A* to *B*, and only the slower ones to pass from *B* to *A*. He will thus, without expenditure of work, raise the temperature of *B* and lower that of *A*, in contradiction to the second law of thermodynamics. [1871:328]

It is difficult to know how seriously Maxwell took this heuristic fiction. Its brevity and location suggest that the passage was almost an afterthought. Nevertheless, Maxwell's Demon generated a debate that even today continues to engage the attention of mathematicians, physicists, and engineers. We may wonder why, since the second law was never seriously in doubt. We may also wonder why the "being" came to be called a "demon," a word Maxwell himself did not use.

These speculations point toward the surplus meaning that the heuristic embodies. The developments that made scientists think of the second law as probabilistic rather than strictly causal were part of the broader movement within physics which culminated in quantum mechanics, which when it abandoned the notion of causality drove Einstein to the riposte that God does not play dice with the universe. What better agent to restore an imperiled causality than a being whom Maxwell described as "essentially finite" (as opposed to an infinite God) and yet "with faculties so sharpened" that he would be "able to do what is at present impossible for us." I conjecture that the Demon has fascinated generations of scientists because he mediated between an eroding belief in universal causality (traditionally derived from God as the prime mover) and the belief in science that caused Laplace, when asked about the role of God in the solar system, to reply that he had no need of that hypothesis.

Maxwell's Demon is a fantasy about an animistic figure who can control dissipation through an exercise of will. The Demon thus occupies the slim margin of escape Kelvin left open when he said that heat death is inevitable if man remains "as at present constituted." Changed just enough to enable him to do "what is impossible for us," the Demon transcends human limits but still remains "essentially finite." The subtexts for this fiction are other fictions, intimately familiar to Victorians, about dissolute heirs who squandered their inheritance and who consequently were subject to dire penalties—in this case, to the heat death prescribed by the second law. From the dissolute heir's point of view the story has a happy ending, for it chronicles the appearance of a "being" who sets things right. From the point of view of science the story is a scandal, for the "being" turns out to be a demon who demands the sacrifice of a scientific truth as the price for his intervention. Like guardians of portals to other realms in ancient myths, the Demon is a liminal figure who stands at a threshold that separates not just slow molecules from fast but an ordered world of will from the disordered world of chaos. On one side is a universe fashioned by divine intention, created for man and responsive to his will; on the other is the inhuman force of increasing entropy, indifferent to man and uncontrollable by human will. No wonder, then, that the debate about the Demon focused on determining how he was and was not like a human being. Clarifying this ambiguity was tantamount to establishing the relation of humanity to an entropic universe.

Other avenues of attack could have been used. For example, Maxwell had supposed the Demon's door was frictionless, so that he could operate it without doing work. Why did this assumption not become a focus of attention? It is true that classical thermodynamics routinely arrived at general theorems by imagining idealized systems that could be modified by variables introduced to account for friction and other complications always present in the real world. The frictionless door was in this tradition. On a deeper level, however, this avenue was not taken because it resolved the paradox too easily. It was too obvious an objection to account for the issues at stake.[10]

Charting the responses to Maxwell's Demon is like mapping the progress of Christopher Columbus across the ocean. From the compass readings we can infer what the prevailing winds and currents were, even though we cannot measure them directly. It is like Columbus's route in another respect as well: only in retrospect does the journey appear as progress toward a certain end. Commentators on Maxwell's Demon tend to treat the responses as if they constitute a narrative of continuous development. This "tradition," however, was largely the creation of writers who searched the scientific literature for predecessors and gave them credit for their own later insights. In histories of Maxwell's Demon, Leo Szilard (1929) is usually presented as having posed important questions that scientists mulled over until Leon Brillouin finally proposed an answer in 1951.[11] But Brillouin's contemporaries were largely unacquainted with Szilard's paper until he cited it as an important contribution to the controversy. It was only *after* Brillouin's citation that Szilard's paper was routinely mentioned. The point is important, for it suggests that although ideas can be conceived at any time, they are converted into a tradition only when they can be synthesized into the prevailing paradigm. In Szilard's case, it was not until computer memory was a possibility that his identification of the Demon's sort-

[10]Later commentators argued that resistance on the door could be made negligibly small; see, for example, Bennett, 1987. My purpose in highlighting this assumption is not to bring this assertion into question but to point out that it was never rigorously tested in the same way that Brillouin, Landauer, and Bennett questioned other aspects of the heuristic.

[11]See, for example, W. Ehrenberg, "Maxwell's Demon," *Scientific American* 217 (1967): 103–110. Ehrenberg's account of the thirty-year gap between the publication of Szilard's paper and its reappearance in Brillouin is typical. "Physicists were preoccupied with so many basic developments that Szilard's postulate was not seriously reviewed until 1951," Ehrenberg writes (p. 109). The assertion is a transparent attempt to create narrative continuity in the face of an obvious gap and shows how the idea of a "tradition" is reinforced through a retrospective reading of history. J. R. Pierce, *Symbols, Signals, and Noise: The Nature and Process of Communication* (New York: Harper & Row, 1961), argues that Szilard's paper had little or nothing to do with the development of information theory, other than its impact on Brillouin (pp. 21–44).

ing as a "kind of memory" was seen as an important clue. When Brillouin happened upon (or sought out) Szilard's paper after computers were well established, this idea allowed him to break the long association between the Demon and human intelligence and conceive the problem from a new perspective.

Szilard had observed that to do his work, the Demon needed to remember where the fast and slow molecules were stored. "If we are not willing to admit that the second law is violated," Szilard wrote, "then we have to conclude that . . . [this kind of memory] is indissolubly connected with the production of entropy" (Ehrenberg, "Maxwell's Demon," p. 109). When Brillouin resurrected Szilard's paper, he refined (or redefined) Szilard's "memory" by identifying it with information. Brillouin pointed out that Maxwell's chamber was a black body, that is, a box whose walls radiate energy at the same frequency as they absorb it. Since vision depends on sensing a difference between absorbed and radiated light frequencies, the demon would have no way to "see" the molecules. Imagine trying to see a black object in an absolutely dark room; that is the Demon's position inside the box.

Then Brillouin demonstrated that if a source of illumination is introduced (a headlamp, for example), the absorption of this radiation by the system increases the system's entropy more than the Demon's sorting decreases it. Thus information gathered by the Demon is "paid for" by an increase in entropy. This result resolves the conundrum by showing that for the system as a whole, the second law is not violated. More important than saving the second law (a quixotic adventure, since it was not in jeopardy) was Brillouin's intuition that entropy and information are connected. This insight led directly to Brillouin's conclusion that information is defined by the corresponding amount of negative entropy (1951).

Just as Brillouin sought out Szilard's paper and retrospectively established it as his predecessor, so his work has been read retrospectively by himself and others. In these readings, the aspect of Brillouin's solution that is underscored is the creation of a context in which information could be divorced from human intelligence. Ehrenberg's article reviewing the controversies over Maxwell's Demon points out that in Brillouin's analysis "the agent does not rely on his intelligence, since he needs physical means to obtain the information—but given the physical means we do not need the agent any longer because we can replace him by a machine!" (p. 109). Brillouin expanded upon this aspect of his interpretation by later emphasizing that information theory completely eliminates the "human element" (1956:x). The threshold was passed; the Demon no longer functioned as a liminal figure mediating between human limitations and inhuman entropy. The dream behind the Demon was realized in another sense, however, for the potent new force of information had entered the arena to combat entropy.

As the implications of Brillouin's analysis were explored, the tensions embedded in the subtext of Maxwell's Demon were transformed rather than resolved. One view, which set human intelligence and will against entropy, became irrelevant; another, which pitted information and machines against entropy, had emerged. From now on, control was increasingly seen not in terms of human will fighting universal dissipation but as information exchanges processed through machines. As the world vaulted into the information age, the limiting factor became the inability of human intelligence to absorb the information that machines could produce.[12]

This current of thought surfaced explicitly in the most recent addition to the heuristic, based on work by Rolf Landauer and Charles H. Bennett (Bennett and Landauer, 1985; Bennett, 1987). Following Brillouin, scientists in the 1950s thought that data-processing operations were thermodynamically irreversible, that is, that heat dissipation was the price that had to be paid for the information-processing operation. Because the amount of heat involved was extremely small, the question was of little practical consequence. But it was important theoretically, for it posited a connection between entropy and information which reinforced Brillouin's interpretation of Maxwell's Demon. When Landauer performed calculations to test this hypothesis, he found that only some data-processing operations were thermodynamically costly. Specifically, he discovered that the irreversible (or costly) processes were those that required the destruction of information (Bennett and Landauer, 1985).

[12]For a discussion of the connections between information technologies and the idea of control, see James R. Beninger, *The Control Revolution: Technological and Economic Origins of the Information Society* (Cambridge: Harvard University Press, 1986).

Armed with this demonstration, Bennett returned to Szilard's paper to confirm that the Demon's operation really consisted of two steps: a measurement step and a memory step (because the Demon has to remember where he put the slow and fast molecules). Brillouin had suggested that the entropic increase was a result of the measurement step. Bennett (1987) refuted this idea by devising a reversible measuring device that allowed measurements to be made without an increase in the entropy. In effect, he showed how the Demon could find out where the molecules were without the need for a headlamp. He then proposed that the true source of the increasing entropy was in the memory step. Since the demon will eventually run out of memory space if he does not clear outdated information from his memory, at some point he must destroy information; and according to Landauer's demonstration, the destruction of information has to be paid for by an increase in entropy. If one supposes a demon (or a computer) with a very large memory, he could of course simply remember all the measurements. The trouble with this scenario, Bennett explains, is that "the cycle would not then be a true cycle: every time around, the engine's memory, initially blank, would acquire another random bit [of information]. The correct thermodynamic interpretation of this situation would be to say that the engine increases the entropy of its memory in order to decrease the entropy of its environment" (p. 116).

Acknowledging that "we do not usually think of information as a liability" (p. 116), Bennett proposes an analogy to make his conclusion plausible. "Intuitively, the demon's record of past actions seems to be a valuable (or at worst a useless) commodity. But for the demon 'yesterday's newspaper' [the result of a previous measurement] takes up valuable space, and the cost of clearing that space neutralizes the benefit the demon derived from the newspaper when it was fresh" (p. 116). Arriving belatedly in the tradition surrounding Maxwell's Demon, Bennett can appreciate more easily than his predecessors that there is a strong correlation between his explanation and his historical moment. He ends his article with the conjecture that perhaps "the increasing awareness of environmental pollution and the information explosion brought on by computers have made the idea that information can have a negative value seem more natural now that it would have seemed earlier in this century" (p. 116).

The surplus meaning characteristic of heuristic fictions thus receives explicit acknowledgment from within the scientific community. Moreover, the play of excess meaning has brought forth interpretations of the heuristic that are concerned with the way information is created and destroyed. The stage is set for a self-reflexive moment to occur.

To see how this moment arrives, consider the multileveled self-reflexive nature of Bennett's interpretation. When he argues that the crux of the problem lies in the destruction of information, his interpretation is engaging in an erasure of previous information analogous to what is happening within the heuristic, according to his interpretation. Of course earlier interpretations also had to compete against received ideas to gain acceptance. But in Bennett's case, the operation performed by his interpretation is mirrored by the operation that he sees the Demon performing. Moreover, he recognizes that his interpretation reinscribes within the heuristic cultural forces operating at the moment of interpretation. Thus the heuristic is seen as a kind of permeable membrane connecting the disciplinary tradition to the culture. Just as the culture is becoming aware that old newspapers do not spontaneously disappear but pile up, so the demon is interpreted as needing to clear "yesterday's newpaper" out of his head. As metaphor is enfolded into metaphor, the scientific tradition is forced to confront the fact that thought, language, and social context evolve together. Social context affects language, language affects thought, thought affects social context. The circle is closed. Objectivity has given way to hermeneutics.

I want now to turn to another juncture in the cascading bifurcations that mark interpretations of Maxwell's Demon. The juncture occurs when Leon Brillouin and Claude Shannon diverge in their opinions about what the relationship between information and entropy should be. In Brillouin's analysis of Maxwell's Demon, the Demon's information allowed him to sort molecules, thus decreasing the system's entropy; but this information had to be paid for by an even greater increase in entropy elsewhere in the system. For Brillouin, then, information and entropy are opposites and should have opposite signs. He emphasized the inverse connection between information and entropy by coining "negentropy" (from negative entropy) as a synonym for information.

To Shannon, an engineer at Bell Laboratories who published a two-part paper that was to form the basis of modern information theory (1948), information and entropy were not opposites. They were identical. When Shannon devised a probability function that he identified with information, he chose to call the quantity calculated by the function the "entropy" of a message. Why he made this choice is unclear. Rumor has it that von Neumann told Shannon to use the word because "no one knows what entropy is, so in a debate you will always have the advantage."[13] One could argue that von Neumann's comment was only one element and that the choice of "entropy" was overdetermined, with multiple factors leading to its conflation with "information." On a conceptual level, an important consideration was the similarity between Shannon's equation for information and Boltzmann's equation for entropy. Because the two equations had similar forms, it was tempting to regard the entities they defined as the same. On the level of language, entropy was compelling because it was a term of recognized importance and could be expected to grant immediate legitimacy to the concept of information. On a cultural level, Shannon's choice anticipated the contemporary insight that proliferating information is associated with the production of entropy. Recall, for example, Landauer's conclusion that it is not obtaining but *erasing* information that dissipates energy. The proposition implies that too much information, piling up at too fast a rate, can lead to increasing disorder rather than order. For postmodern society the compelling fable is not Maxwell's Demon but *My Brother's Keeper*, the story told by Marcia Davenport (1954) about two reclusive brothers who were finally buried under the copies of the *New York Times* they compulsively saved. What we fear most immediately is not that the universe will run down, but that the information will pile up until it overwhelms our ability to understand it.

[13]This version comes from Campbell, 1982:32. Slightly different phrasing is cited in Denbigh and Denbigh, *Entropy*, p. 104; they give their source as Myron Tribus, *Boelter Anniversary Volume* (New York: McGraw-Hill, 1963).

Whatever the reasons for Shannon's choice, it is regarded by many commentators within the scientific tradition as a scandal, for it led to the (metaphoric) knotting together of concepts that are partly similar and partly dissimilar. Typical is K. G. and J. S. Denbigh's reaction in their careful study of the way the quantity defined by Shannon's equation differs from thermodynamic entropy. Recounting the story about von Neumann's advice, they write that thus "confusion entered in and von Neumann had done science a disservice!"[14] Jeffrey S. Wicken is even more explicit, calling Shannon's choice "loose language" that served "the dark god of obfuscation." "As a result of its independent lines of development in thermodynamics and information theory, there are in science today two 'entropies,'" Wicken writes. "This is one too many. It is not science's habit to affix the same name to different concepts. Shared names suggest shared meanings, and the connotative field of the old tends inevitably to intrude on the denotative terrain of the new" (1987:183).

Clearly Wicken's concern is to restore scientific univocality by closing off the ability of the information-entropy connection to act as a metaphor rather than a congruence. Yet at the same time he admits that shared language creates an inevitable "intrusion" into the "denotative terrain" of one term by the "connotative field" of another. The problem is more scandalous than he recognizes, for whenever a heuristic is proposed, it necessarily uses "shared names" that cause scientific denotation to be interpenetrated by cultural connotations. For what else is language but "shared names"? As Wittgenstein has observed, there are no private languages. Moreover, the distinction between denotative and connotative language is itself part of the distinction between language-as-vehicle and language-as-concept which metaphors, and particularly self-reflexive metaphors, bring into question. To turn Wicken's argument on its head, we might say he recognizes that metaphors in general, and the information-entropy connection in particular, directly threaten science's ability to separate ideas from the language it uses to express them.

In his anxiety to suppress the metaphoric potential of Shannon's choice, Wicken misses the richly complex and suggestive connections

[14]Denbigh and Denbigh, *Entropy*, p. 104.

that were instrumental in enabling a new view of chaos to emerge.[15] By the simple device of using "information" and "entropy" as if they were interchangeable terms, Shannon's choice gave rise to decades of interpretive commentary that sought to explain why information should be identified with disorder rather than order. For the alliance between entropy and information to be effective, information first had to be divorced from meaning (a premise made explicit in Shannon's 1948 papers) and had to be associated instead with novelty. Recall the random number generator, mentioned earlier, that produces a tape we can read. No matter how long we watch the tape, numbers keep appearing in unpredictable sequence. From one point of view this situation represents chaos; from another, maximum information.

Once randomness was understood as maximum information, it was possible to envision chaos (as Robert Shaw does) as the source of all that is new in the world.[16] Wicken is correct in noting that denotative and connotative fields overlap; in the case of information, the connotation that "intruded" upon the denotative field of chaos was complexity. Whereas chaos had traditionally meant simply disorder, complexity implied a mingling of symmetry with asymmetry, predictable periodicity with unpredictable variation. As we have seen, chaotic or complex systems are disordered in the sense that they are unpredictable, but they are ordered in the sense that they possess recursive symmetries that almost, but not quite, replicate themselves over time. The metaphoric joining of entropy and information was instrumental in bringing about these developments, for it allowed complexity to be seen as rich in information rather than deficient in order.

To see how Shannon's choice embodies a self-reflexive moment, it will be necessary to understand more precisely how informational

[15] Wicken, 1987, cites some of the order-out-of-chaos work, but he sees these developments as occurring despite the ambiguity in informational entropy rather than as being facilitated by it. Drawing from this work, Wicken proposes that what Shannon's function actually measures is the "complexity" of a message (p. 184). Wicken suggests that the use of this term instead of "entropy" would remove confusion about whether information is or is not ordered.

[16] Shaw, 1981. Of contemporary chaologists, Shaw is perhaps the best known for positing and developing a strong connection between chaos and information.

entropy is like and unlike thermodynamic entropy. Shannon defined information as a function of the probability distribution of the message elements.[17] Information in Shannon's sense does not exist in the same way as the dimensions of this book exist. A book can be measured as twelve inches long, even if there are no other books in the world. But the *probability* that a book has that dimension is meaningful only if there are other books with which it can be compared. If all books are twelve inches long, the probability that a given book has that dimension is 1, indicating complete certainty about the result. If half of the books are twelve inches, the probability is 1/2; if none are, it is 0. Similarly, information cannot be calculated for a message in isolation. It has meaning only with respect to an ensemble of possible messages.

Shannon's equation for information calculated it in such a way as to have it depend both on how *probable* an element is and on how *improbable* it is. Having information depend on the probability of message elements makes sense from an engineer's point of view. Efficient coding reserves the shortest code for the most likely elements (for example, the letter *e* in English), leaving longer codes for the unlikely ones (for instance, *x* and *z*). Improbable elements will occupy the most room in the transmission channel because they carry the longest codes. Thus for a channel of given capacity, fewer improbable elements can be sent in a unit of time than probable ones. This explains why an engineer would think it desirable to have a direct correlation between probability and information.

Why have information correlate with improbability? Partly compensating for the longer codes of improbable elements is the greater information they carry. To see why improbable elements carry more information, suppose that I ask you to guess the missing letter in "ax—." It is of course *e*, the most probable letter in an English text. Because it is so common, *e* can often be omitted and the word will still be intelligible. In "axe," the letter *e* carries so little information that "ax" is an alternate spelling. Suppose, by contrast, that I ask

[17]Shannon's equation calculated the information H as $H = -\sum_{i=1}^{n} P_i (\log_2 P_i)$, where n is the number of different kinds of symbols that could be used in the message (for example, 26 alphabet letters) and P_i is the probability of the ith kind.

you to guess the word "a—e." You might make several guesses without hitting the choice I had in mind—"ace," "ale," "ape," "are," "ate." When you find out that the expected letter is *x*, you will gain more information than you did when you learned that the final letter was *e*. Shannon's equation recognizes this correspondence by having the information content of a message increase as elements become more improbable.

This dual aspect of information is immediately apparent when information is plotted as a function of probability. The resulting curve is a parabolic arc, as shown in figure 1. (This diagram describes the simplest case, when the probabilities of message elements are independent of one another.) The diagram indicates that as the probability increases, the information increases until it reaches a maximum when the probability is 1/2. Then it begins to decrease as the message becomes highly probable. When the probability is 1—that is, when there is no uncertainty about what the message will be—it drops to zero, just as it does when the message is completely improbable. Maximum information is conveyed when there is a mixture of order and surprise, when the message is partly anticipated and partly surprising.

We are now in a position to understand exactly how Shannon's equation for informational entropy differs from thermodynamic entropy. Although Shannon's equation for information has the same form as Boltzmann's equation for entropy, the *meanings* of the probabilities differ in the two equations. In Boltzmann's equation, the probabilities derive from a lack of specific information about the system's microstates. Thermodynamic quantities such as tempera-

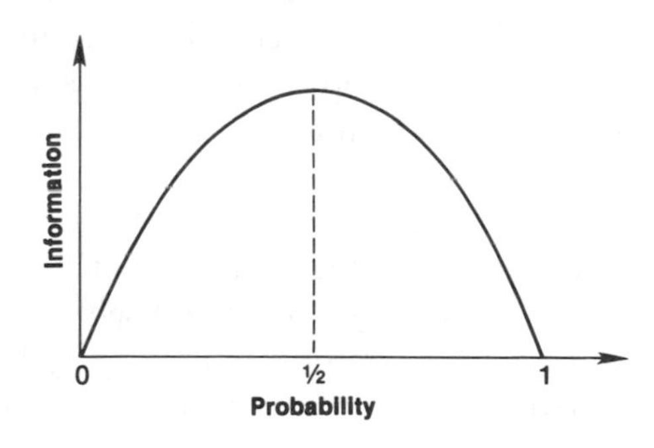

1. Information as a function of probability

ture and entropy are macrostate properties; that is, they are statistical averages that represent the collective actions of millions of molecules. According to the uncertainty principle, we have no way to know how a single particle behaves. Our ignorance of the microstates is reflected in the probabilistic form of Boltzmann's equation. In Shannon's equation, by contrast, the probabilities derive from *choice* rather than ignorance; they reflect how probable it is that we would choose one message element rather than another, given a known ensemble (for example, the alphabet).

Wicken has a good example that can be adapted to illustrate this difference (1987:185). Suppose that I ask you to place a die on a table a thousand consecutive times, each time choosing the face you want to place upward. Shannon's equation calculates the information on the resulting sequence as a function of the number of times a face actually turns up in comparison with the number of times it could be predicted to turn up, given the number of faces on the die. The uncertainty implied by the probability function reflects my inability to know in advance which choices you will make, not my ignorance about which die faces have already appeared. To imagine an analogous case for thermodynamic entropy, suppose that a thousand dice are cast all at once, and a measuring instrument records the total amount of light reflected from the die faces. We do not know how each individual die landed. After calibrating the light instrument, however, we could figure out the average face count on the basis of the amount of light reflected. In this case the probability function reflects ignorance of the microstates, not ignorance of our choices in assembling a series of such states.

The heuristic that evolved around Shannon's choice does not make clear the difference between informational entropy and thermodynamic entropy. On the contrary, since the point of the heuristic is to justify Shannon's choice, differences between informational and thermodynamic entropy are suppressed in favor of similarities. But as in the case of metaphors, the differences are not negligible. They put a torque on the heuristic which twists the way entropy is understood. This torque registers itself as a perturbation in the language of the commentators, as they struggle to suture the gap created by difference without ever quite acknowledging that there *is* a gap.

Warren Weaver was Shannon's first, and perhaps most important, commentator.[18] His explanation for the correlation of information with disorder rather than order set a precedent that other commentators would follow for at least twenty years. Reasoning that if a message is perfectly ordered, the receiver will be able to guess what it will say, Weaver suggests that a "noisy" message will be more surprising and hence will convey more information. He is now in a quandary, for by this reasoning gibberish should convey the maximum possible information. To close off this possibility, Weaver introduces a distinction between desirable and useless information. True, gibberish is maximum information. But since it is not desired, it does not really count as information. Hence the maximum amount of information is conveyed by a message that is partly surprising and partly anticipated.

This explanation produces the result required by the theory, which, as we have seen, defines information through a curve that reaches its maximum precisely at the midpoint between certainty and uncertainty. To arrive at this conclusion, the heuristic must inject the receiver's knowledge as a factor. But the probabilities defined by Shannon's equation do not depend on the receiver's knowledge of the message. They are determined solely by the frequency with which a given element appears in relation to its predicted frequency of appearance in the ensemble. To suture this gap, Weaver turned his attention to a quantity in one of Shannon's equations which did depend on the receiver's knowledge—an ambiguous quantity that Shannon called, appropriately, the equivocation.

One of Shannon's important contributions was to create a schematic of the communication situation which made clear that there is no such thing as an unmediated message. By dividing the communication situation into a sender, an encoder, a channel, a decoder, and a receiver, Shannon demonstrated that any message is always subject to the intrusion of "noise." Noise can be anything that interferes with the reception of the message the sender sent—misprints in a book, lines in a TV image, static on a radio, coding errors in a telegram, mispronunciations in speech. Noise is measured in the same units as information; indeed, it *is* information, but information not

[18]Warren Weaver's commentary, which first appeared in *Scientific American*, was bound together with Shannon's two papers in Shannon and Weaver, 1949.

intended by the sender. The amount of information contributed by noise is the "equivocation."

As an employee of AT&T, Shannon was interested in transmitting messages as accurately as possible, and he naturally considered the equivocation to be an unwanted intrusion that should be subtracted from the received message so that the original message could be retrieved. But Weaver, in his commentary on Shannon's papers, proposed that in some instances the equivocation might be seen as a desirable addition to the message rather than as an interference. This ambiguity in the sign of the equivocation turned out to be extremely fruitful, for it led to a new view of the communication process in which noise was seen as playing a constructive rather than a destructive role.

Henri Atlan's article "On a Formal Definition of Organization" (1974) is based on this view. Atlan points out that equivocation in a message can sometimes lead the system to reorganize itself at a higher level of complexity, as when a genetic mutation results in an adaptive trait. He therefore proposes that we distinguish between two kinds of equivocation—a "destructive" one that interferes negatively with a message and an "autonomy-producing" one that stimulates a system to undergo reorganization. How an "autonomy-producing" equivocation is seen depends on where the observer is stationed. If she is inside the channel, the equivocation is an interference, for within this frame of reference one is interested only in the message. If she is outside the channel, however, she can see the effect on the system as a whole. The observer's knowledge thus reenters the picture, but it is constituted differently than in the Shannon-Weaver heuristic. For Atlan it is knowledge of the system as a whole that counts, not knowledge of the message.

The point I want to make in tracing the genealogy from Shannon's theory to Weaver's commentary to Atlan's proposal is that it was precisely the multivocality of the information-entropy connection that allowed new views to emerge. At the center of this multivocality is a self-reflexive moment. When the equivocation came to be seen as a potentially positive quantity within the Shannon heuristic, the heuristic became a metaphor of itself, for in making equivocation an equivocal quantity, the heuristic acknowledged that there was surplus meaning not only within the communication channel but within itself also. The constructive role that surplus meaning can play was then metaphorically incorporated into the order-out-of-chaos paradigm in the recognition that noise can sometimes cause a system to reorganize at a higher level of complexity. Thus, as with Maxwell's Demon, my story ends with a heuristic that has become self-reflexive on multiple levels.

It remains to clarify what role disciplinary practices had in shaping the heuristics of my story, as well as informing my story of the heuristics. I can illustrate by returning to an unresolved crux in my story—the disagreement between Shannon and Brillouin on whether information should have the same sign as entropy or an opposing sign. When I surveyed several dozen textbooks on information theory to see how they treated the information/entropy crux, I found a clear division along disciplinary lines. Almost without exception, textbooks written by electrical engineers followed the Shannon-Weaver heuristic, explaining that the more uncertain a message was, the more information it could convey.[19] Like Weaver, these writers withdrew from the obvious conclusion that gibberish is maximum information by saying that a mixture of surprise and certainty was needed. Also like Weaver, they did not recognize the implicit contradiction with Shannon's theory. On the whole, they did not devote much space to the relationship of thermodynamic entropy and informational entropy.

Textbooks written by chemists, physicists, and thermodynamicists, by contrast, usually adopted the Brillouin explanation, developing the concept of information through its connection with thermodynamic entropy.[20] Maxwell's Demon figured prominently in these explanations and led to the expected conclusion that entropy and information should have opposite signs. For these authors the

[19]Typical of commentators who follow the Shannon-Weaver heuristic is Gordon Raisbeck, *Information Theory: An Introduction for Scientists and Engineers* (Cambridge: MIT Press, 1964), pp. 1–11 especially. Raisbeck is steeped in the engineering tradition; he was Norbert Wiener's son-in-law.

[20]For an example of the Brillouin tradition, see D. A. Bell, *Information Theory and Its Engineering Applications* (London: Pitman, 1953), especially pp. v and 120–125. Bell's discussion is noteworthy because he considers untangling the relation between entropy and information one of the important problems his book addresses. Sometimes hybrid explanations result, as when Jagjit Singh (1966) mixes a microstate/macrostate analogy with the Shannonian uncertainty in the message argument.

problem of how thermodynamic entropy related to informational entropy was compelling; most of them devoted several pages to the question. The ability of both heuristics to replicate themselves through several generations of textbooks is striking evidence of the effectiveness of disciplinary traditions in erecting boundaries that marginalize or trivialize what happens outside of them.

In fact, the problem of how Brillouin's negentropy relates to Shannon's entropy is not especially complex or difficult. More than twenty years ago, John Arthur Wilson (1968) demonstrated that Brillouin's proofs still hold true if the signs are reversed and the "negentropy" concept is dropped. But the debate continues because the heuristics are informed by other associations. Brillouin's heuristic grew out of his analysis of Maxwell's Demon, and this analysis makes sense *only* if information and entropy are opposites. Shannon's heuristic, by contrast, concentrates on the circuits necessary to transmit messages, and these circuits emphasize the intrinsic uncertainty of message transmissions. Embodied in the heuristics are values extraneous to the formal theories but essential to the mindsets out of which they grew.

The crucial differences revealed by the heuristics are two opposite ways of valuing disorder. These differences are implicit in Wilson's reasons for dropping "negentropy." He argues that by defining information as the opposite of entropy, Brillouin is "attributing to information a quality which it does not have—the quality of being organized" (p. 535). It is the connotation that disorder is the enemy of order and thus of information that is debated in the heuristics, not the denotative results of the theories themselves. When information could be conceived of as *allied* with disorder, a passage was opened into the new paradigm of chaos theory.

Within the heuristics, the crux of the disagreement lies in where the commentator positions himself with respect to the transmission process. Both heuristics agree that entropy correlates with uncertainty. But the Shannon heuristic foregrounds the uncertainty present before the message is sent, whereas the Brillouin heuristic focuses on the uncertainty that remains after the message has been received. The difference between the two viewpoints can now be succinctly stated. *Shannon considers the uncertainty in the message at its source, whereas Brillouin considers it at the destination.* To ask which is correct is like asking whether a glass is half empty or half full. The answer is important not because it is correct but because it reveals an orientation toward the glass and, by implication, an attitude toward life. Similarly, the Brillouin and Shannon heuristics reveal different attitudes toward chaos by their orientations toward the message.

Like the optimist and pessimist regarding a glass of water, Shannon and Brillouin locate themselves at the halfway point of the information-probability arc and look in opposite directions, as in figure 2. Shannon, looking forward, sees a downward-sloping curve and argues that the more certain the message is, the less information it conveys. Brillouin, looking backward, also sees a downward curve and argues that the more surprising a message is, the less information it conveys. Both recognize that maximum information comes when there is a mixture of certainty and surprise. But where Brillouin emphasizes certainty, Shannon stresses surprise.

These different orientations are no doubt related to the different contexts in which the two men lived and worked. Brillouin began his career in thermodynamics, a discipline that had traditionally envisioned disorder as the enemy. From Kelvin on, thermodynamicists had seen entropy as an inhuman chaos that would win in the end, try as one might to resist it. Shannon worked for AT&T, a company that made its living by satisfying people's curiosity. The more uncertain people were—about the stock market, national news, or events in other cities—the more telegrams they sent, phone calls they made, information they required. No wonder Shannon thought of uncertainty as information's ally, whereas Brillouin saw the two as antag-

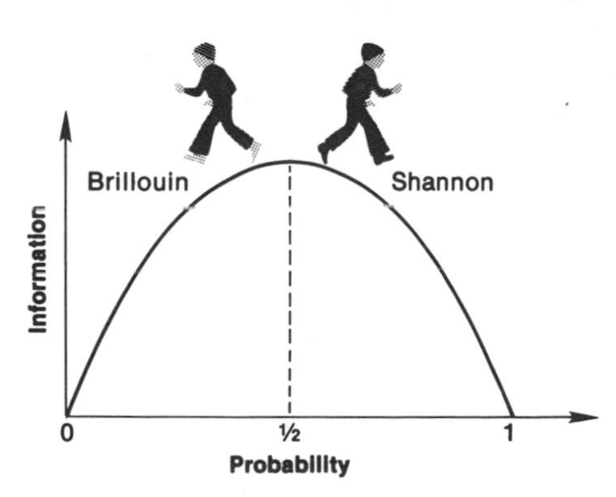

2. Locating Shannon and Brillouin on the information/probability curve

onistic. The controversy illustrates the fact that scientific traditions are seen in a false light if they are regarded solely as self-contained, rationally coherent discourses. Concepts and theories are important. But so are the heuristics that explain them, the languages that constitute them, the disciplinary practices that inform them, and the cultural contexts that interpenetrate them.

Within the heuristic I have created to tell my story—that is, the self-reflexive moment—the values of my own discipline are clearly inscribed. For someone steeped in literary analysis, it is a given that multiple signification is a plus rather than a minus, or to use metaphors more appropriate to literature, a story rather than a scandal. The metaphoric play that I have been excavating throughout this chapter leads me to a different interpretation of Borges's image of a stairway ending in space. Although it does not lead to the expected door, it can sometimes turn into a passage that opens onto previously unrecognized and unconstituted territory. As we shall see in the next two chapters, disorder is not necessarily bad, and the void is not always empty.

CHAPTER 3

The Necessary Gap: Chaos as Self in *The Education of Henry Adams*

IN scientific texts the connections among concept, culture, and self are implicit, encoded in metaphors and heuristics. In *The Education of Henry Adams* they are explicitly the subject of Adams's dynamic theory of history. According to the dynamic theory, the curve of history can be charted as the attraction between the human mind and the explosive forces of a supersensual chaos. Modern consciousness differs from earlier modes of thought because it has absorbed more of the chaos within itself. In *The Education*, chaos is not just connected with the self; in an important sense, it *is* the self. Whereas my discussion of the scientific tradition started with objectivity and moved toward subjectivity, my comments about *The Education* will start with how subjectivity is constituted within the text and move toward its entanglement with scientific concepts.

Privately printed by Adams in 1907, *The Education* marks a midpoint between Kelvin's mid-nineteenth-century vision of entropy as universal dissipation and Shannon's reconceptualization of it as information in the mid–twentieth century. Anticipating Shannon, *The Education* recognizes the possibility that chaos may be positive force rather than negative uncreation. But enough of the older attitudes toward disorder still cling to it so that chaos is represented within *The Education* as an intensely ambivalent, as well as highly charged, concept. Even whether it is something rather than nothing, a presence rather than an absence, remains in doubt. This doubt is reflected and reinscribed in the constitution of a "real" self, which paradoxically is a presence constituted through the assertion of its absence. Signifiers of the self include gaps, holes, and voids in the narrative. In the resulting interplay between presence and absence, meaning derives as much from what is not said as from what is said. "The result of a year's work depends more on what is struck out than on what is left in," Adams wrote in *The Education*. It also depends more "on the sequence of the main lines of thought, than on their play and variety" (1973: 389).

According to what has been left in, the sequence appears to be this. Adams self-consciously conceived of *The Education* as marking the rupture between the ordered certainties of the Newtonian synthesis and the chaotic multiplicities that he saw as characteristic of the twentieth century. This division is inscribed into the form of *The Education*. The first half records Adams's repeated attempts to launch himself in the world, working from a conception of the universe as unity, linearity, and fixed truths; the second half finds him searching for, and eventually articulating, a theory that can explain the world as it actually exists—an anarchistic multiverse of chaos, complexity, and relativism. From this perspective, *The Education* seems to be an exemplary account of one man's initiation into the technological and social contexts that form the cultural background for the later emergence of the sciences of complexity.

Yet almost immediately, complications arise in this simple picture. Henry Adams himself is not a unit but a multiple. There is the character whom Adams describes in the Preface as a "manikin," of interest only because of the clothes he wears (p. xxx); the narrator, who writes retrospectively about Adams the character; and the author, who constructs both narrator and character as well as the clothes they wear.[1] If we choose to omit the author and arrange the narrator and character along a linear sequence, we can say (as various commentators have said) that the gap between the naive perspective of the character and the educated hindsight of the narrator continually

[1]The division into author, character, and narrator is conventional enough to be often noticed; see, for example, Martin, 1981:119. The clothes reference is of course an allusion to Carlyle's *Sartor Resartus*, to which Adams alludes throughout *The Education*.

narrows as the text progresses, until by the end the character espouses a view very close to the narrator's.[2] By the nature of textuality, however, the gap never quite closes. As the presumed present-tense existence of the narrator is translated into the past tense of the character, the character stands as a marker for what was, forever deferred and different from the narrator.

This complication is itself a simplification; but it is nevertheless useful, for it demonstrates how far the text is from a merely linear sequence. The narrator's retrospective view creates an undertow that pulls against the character's naiveté, transforming the narrative into a strongly nonlinear progression. On the opening page, for example, after a circuitous sentence giving the place and time of his birth, Adams muses on whether his distinctive heritage was an advantage or a handicap. Characteristically he finds himself, even in retrospect, unable to decide. He knows only that he "could not refuse to play his excellent hand." He asserts that he "accepted the situation as though he had been a party to it, and under the same circumstances would do it again, the more readily for knowing the exact values." Apparently the values are not so exact, however, for within a few lines he remarks that "as it happened, he never got to the point of playing the game at all; he lost himself in the study of it, watching the errors of the players" (p. 4).

If we think of the narrator as an extrapolation into the present of Adams the character, there is an obvious contradiction in his claims that Adams did not refuse to play and that he only observed. The contradiction is intensified by the qualifying phrase "as it happened," for the phrase insists on the actuality of the narrator's observer status, even as it grounds itself in a historicity appropriate to the character's player status. The construction implies that in some sense the narrator and character are at once separated in time and simultaneously present, part of a linear sequence and enfolded together into a convoluted space that defies linearity. Given this space, it is simplistic to the point of distortion to say that *The Education* records Adams's initiation into chaos. Rather, chaos is represented

[2]Judith Shklar discusses the narrowing gap between narrator and character in "*The Education of Henry Adams* by Henry Adams," *Daedalus* 103 (1974): 59–66, especially p. 61.

as always already present within a complex dynamic of revelation and concealment.

In an often-cited letter to Henry James, Adams called *The Education* "a mere shield of protection in the grave," adding, "I advise you to take your own life in the same way, in order to prevent biographers from taking it in theirs."[3] The remark suggests that as one life is scripted into the text, another is leached away. But Adams does not imagine himself dying *into* the finished text, as Nabokov's protagonist Van Veen claims in *Ada*. On the contrary, the inscripting process is to provide a shield, a protective covering that conceals even as it marks a rupture between Adams the author and the character/narrator of *The Education*. What does this rupture signify? On the surface, it makes the same point Adams stresses in the Preface, that the character we see in *The Education* is no self-revelation of an egoistic individual but a manikin who has the "same value as any other geometrical figure of three or more dimensions, which is used for the study of relation" (p. xxx). More subtly, the rupture constitutes as a possibility the belief that some authorial self lingers beyond the reach of textuality, autonomous and self-determined in a way that the character obviously is not. Thus the very act of concealing a "real" self is also a constituting gesture that establishes its existence. It is in this sense that the absence—or more correctly, the suppression—of self affects the form of the text as much as if its presence were emblazoned on every page.[4]

The "real" self, manifesting itself within the text as absence, rupture, or gap, further complicates the linear flow of the narrative and

[3]Dated 6 May 1908, the letter appears in Adams, 1938:495.

[4]Candace Lang, in "Autobiography in the Aftermath of Romanticism," *Diacritics* 12 (1982): 1–12, cogently criticizes recent theoretical treatments of autobiography for assuming that an authorial self, however problematic its relation to an autobiographical text, exists outside the text, independent of writing and textuality. She argues that such essentialist notions of an authorial self are naive, because the self is always already textualized in the Derridean sense. I accept this view. In writing about a "real" self, I mean to imply not a self that is essential in fact but one posited *within the text* as beyond the reach of textuality. That such a self is itself a fabrication of the text is assumed as self-evident. For a reading of *The Education* that applies deconstructive premises to Adams's construction of a self, see John Carlos Rowe's provocative study *Henry Adams and Henry James* (1976). Rowe asserts that in "the context of Adams's thought, the conventional notions of 'author' and 'subject' cannot be sustained" (p. 122).

punctuates the accretion of the inscripted self, rendering its evolution discontinuous or indeterminate. Perhaps the chapter that illustrates this process most clearly is "Darwinism." In his attempt to substantiate Darwinist claims for unbroken continuity within evolution, Adams confronts his own evolution as an individual. He discovers that the fossil record used to support Darwin's theory is incomplete and sometimes contradictory. Working strictly from the evidence, he concludes that he "could prove only Evolution that did not evolve; Uniformity that was not uniform; and Selection that did not select" (p. 231). Forced by this lack of evidence to ask why he wanted to defend Darwinism in the first place, he decides that he "was a Darwinian for fun" (p. 232), and that as soon as another new idea arrived, he would "drop off from Darwinism like a monkey from a perch." Embedded in this play of wit is a sobering conclusion—that the "idea of one Form, Law, Order, or Sequence had no more value for him than the idea of none; that what he valued most was Motion, and that what attracted his mind was Change" (p. 231). Since Adams has until this point represented himself as the champion of Law, Order, and Sequence, the conclusion that his mind is attracted by change seems to call for a radical reassessment of who he is.

Instead we are presented with a suture within the text which joins Adams's past and present selves. Significantly, the sequence reestablished by this suturing is not a smooth continuum. Although the suture succeeds in stitching past to present, it also leaves a visible scar, testimony to the depth of the rupture. What caused the rupture we can only infer, since it lies in shadow, hinted at rather than directly stated. Nevertheless, it proves to be a powerfully attractive force for Adams. "Psychology was to him a new study and a dark corner of education. As he lay on Wenlock Edge, with the sheep nibbling the grass close about him as they or their betters had nibbled the grass—or whatever there was to nibble—in the Silurian kingdom of Pteraspis, he seemed to have fallen on an evolution far more wonderful than that of fishes" (p. 231).

Confronting the possibility that the multiplicity he has always fought against is not just an external force but an internal one as well, he decides that he "did not like it; he could not account for it; and he determined to stop it." He fortifies himself with the idea that never "since the days of his *Limulus* ancestry had any of his ascendents thought thus. . . . Out of his millions and millions of ancestors, back to the Cambrian mollusks, every one had probably lived and died in the illusion of Truths which did not amuse him, and which had never changed." When this unbroken chain of antecedents had practiced resistance, it was with the understanding that values were fixed points. They might be wrong, but they never doubted there was an ultimate truth. Into this continuous sequence, stretching to and beyond the prehistoric horizon, a schism erupts in the form of Adams. "Henry Adams was the first in an infinite series to discover and admit to himself that he really did not care whether truth was, or was not, true. He did not even care that it should be proved true, unless the process was new and amusing" (pp. 231–232).

Now comes the suture. Faced with the recognition that he himself is the catastrophe that disrupts sequence and gives the lie to continuous evolution, he is determined to fold under this multiplicitous self that has momentarily emerged. "He had no notion of letting the currents of his action be turned awry by this form of conscience. To him, the current of his time was to be his current, lead where it might. He put psychology under lock and key; he insisted on maintaining his absolute standards; on aiming at ultimate Unity." Although his quest for unity will continue, the narrator now believes that it derives not from innate nature but from will. "The Church was gone, and Duty was dim, but Will should take its place, founded deeply in interest and law" (p. 232). Where does this will originate? Perhaps in Adams the character, as he struggles to merge the current of his thought with the current of his time; perhaps in Adams the narrator, concerned to stitch together the bits and pieces of his recollections into recognizable garments (how arbitrary these seams can sometimes be is demonstrated in the next paragraph, when the narrator turns from currents to currency); or perhaps even in Adams the author, for the suture hides from view a psychology at odds with both character and narrator, thus simultaneously creating a third self and projecting it beyond the reach of textuality.

As I reread the chapter, realizing that the suture exists, I am struck by the deviousness of this text. For on the opening page, I find Adams declaring essentially the same sentiments that later provide material for a crisis. "The ideas [of Darwinism] were new and seemed to lead somewhere—to some great generalization which would finish one's clamor to be educated. That a beginner should understand them all, or believe them all, no one could expect, still less exact. Henry Adams was Darwinist because it was easier than not. . . . He was ready to become anything but quiet. As though the world had not been enough upset in his time, he was eager to see it upset more." Moreover, the narrator admits that Adams "cared nothing about Selection, unless perhaps for the indirect amusement of upsetting curates" (pp. 224–225). Why do these lines have such a different effect than the later realization that he "was a Darwinian for fun"? Partly because they come as prologue rather than conclusion. Adams is a "beginner," and although he is "eager to see [the world] upset more," his desire is presumably on the side of scientific truth rather than against it. Partly too because there is the strong suggestion that his foray into Darwinism is part of a larger approach to truth, or at least to "some great realization that would finish one's clamor to be educated." When the sequence leads to rupture rather than closure, one is left with the impression that something out of the ordinary has happened, something Adams could not reasonably have expected. And yet when I return to the chapter's beginning with foreknowledge of the end, I can see plainly enough that it was all there in the beginning, had I only known how to read it. Which is, of course, precisely the perspective from which the narrator writes.

The chapter demonstrates, then, that the text is informed by two systems of encoding—one embodying the forward, linear perspective of the character, the other giving the backward, retrospective view of the narrator. For critics, this double encoding makes argument by quotation hazardous in the extreme. Depending on the context in which quotations are embedded, one set of codes can be suppressed in favor of another, so that the same words can have not just different but opposite import. Add the further complexities introduced by ruptures, and one can sympathize with the narrator's complaint that no one "means all he says, and yet very few say all they mean, for words are slippery and thought is viscous" (p. 451). In the face of these uncertainties, commentators on *The Education* have looked for ways to stabilize their interpretations by concentrating on large structural blocks or, if they remain on the level of sentences, by looking for consistent patterns repeated throughout the text. Yet even here one's grasp on the results is precarious, for in a rhetoric so full of abrupt turns, it is easy to shoot off in one direction while the text turns in another.

Yet the turns themselves form a pattern. *The Education* is testimony to Adams's exquisite sense of form, as he comes close to admitting in one of the rare passages in *The Education* where he comments specifically on his writing. "The pen works for itself, and acts like a hand, modeling the plastic material over and over again to the form that suits it best. The form is never arbitrary, but is a sort of growth like crystallization, as any artist knows too well; for often the pencil or pen runs into side-paths and shapelessness, loses its relations, stops or is bogged. Then it has to return on its trail, and recover, if it can, its line of force" (p. 389). Following the maxim "form is never arbitrary," I want to sketch out a pattern that generalizes the observations I have made so far. I shall proceed by way of two commentaries on *The Education*, one structural and one syntactic.

The first is James M. Cox's article "Learning through Ignorance: *The Education of Henry Adams*" (1980). Recalling that a dynamo is constructed of two magnetic poles across which a current arcs, Cox points out that bipolarity is everywhere in Adams's work. It is inscribed, for example, in the old and new towers of Chartres that Adams analyzes in *Mont-Saint-Michel and Chartres* (1913); it appears again in the contrast between the femininity of the Virgin's cathedral at Chartres and the masculinity of Saint-Michel; it is written into the relation between *Mont-Saint-Michel and Chartres* as a study of twelfth-century unity and *The Education* as an exploration of twentieth-century multiplicity; and it is inscribed again in *The Education*, with its bifurcated structure broken in half by the twenty-year gap intervening between "Failure" and "Twenty Years After." Cox suggests that we think of *The Education* as a kind of textual dynamo, with a current crackling back and forth across the gap in a dynamic connection that defeats any static or linear conception of the narrative. Although Cox is not concerned specifically with the relation between character and narrator, one could extend the image to them. At the same time that the narrator's retrospective knowledge surrounds the character's naiveté and gives it an ironic inflection, the character's ignorance of outcomes interpenetrates the narrator's knowledge, imparting to it a dynamic uncertainty that belies his posterior position.

Without acknowledging the work of Cox, Linda A. Westervelt provides an important corollary in her analysis of sentence construction in *The Education* ("Henry Adams and the Education of His Readers," 1984). She points out that key passages are often constructed as compound complex sentences, with independent clauses connected by "but." Westervelt argues that the two halves of these sentences do not cancel each other; rather they create a dynamic tension, making it impossible to decide which assertion is correct. On this view the dynamo is something more than an object represented within *The Education*, something more even than the expansive metaphor it becomes in Adams's dynamic theory of history. It is the motor that drives the narrative, from the microcosmic level of individual sentences to the macrocosmic level of the large structural blocks. Combining the work of Cox and Westervelt, one could argue that in the dynamo Adams found a structure, a syntax, and a paradigm all at once.[5]

A strand in Westervelt's work, however, goes beyond and qualifies this picture. The balanced antitheses of the sentence structure create an unpredictability that makes it impossible to predict the sequence of thought. If a linear extrapolation appears likely or certain, it is nearly always a setup, an invitation to catastrophe.[6] We are told, for example, that Clarence King, with "ordinary luck," would "die at eighty the richest and most many-sided genius of his day" (p.

[5]It has been pointed out to me, by a reader who evaluated this book for Cornell University Press, that a dynamo can act either as a generator or as a motor, depending on whether mechanical energy is converted to electricity or electricity to mechanical energy. Adams does not explicitly comment on this duality, but it is entirely appropriate to his ambiguous construction of the text as a dynamo in which it is uncertain in which direction the lines of force run.

[6]David Marcell, in "Henry Adams's Historical Paradigm: A Reexamination of the Major Phase" (in Hague, 1964:127–141), comments that this predictable disruption of sequence is itself a unifying principle. "*The Education* itself becomes unified, paradoxically, by this recurring pattern of logical assumptions and expectations rudely shocked by the caprice of events and the rush of experience" (p. 135). He also argues that the "only unity left was consciousness. . . . Adams had, in fact, been compelled to write his autobiography out of the need to explore the nature of consciousness and its relation to history" (p. 136).

313). Later we learn he in fact died a pauper, ignored and scorned by all but a few loyal friends. So pervasive are discontinuities of this kind that one of the narrator's most characteristic activities is suturing, trying to stitch together a past and a present that have been torn apart and can be fitted together only with difficulty.

At moments of transition, the admission that suturing is necessary comes close to the surface. After the end of the Civil War, for example, the narrator remarks that when Adams "sat down again at his desk in Portland Place before a mass of copy in arrears, he saw before him a world so changed as to be beyond connection with the past" (p. 209). England itself is in need of suturing, for it "was a social kingdom whose social coinage had no currency elsewhere" (p. 203). The cream of English society "could teach little worth learning, for their tastes were antiquated and their knowledge was ignorance to the next generation" (p. 202). English arts and letters were no better. According to the narrator, English thought was as chaotic as the English drawing room, consisting of "bits and fragments of incoherent furnitures which were never meant to go together, and could be arranged in any relation without making a whole, except by the square room" (p. 212). English art was merely a "garden of innate disorder called taste" (p. 214). America was, if possible, worse. When Adams returns to his native land, he decides that one "profession alone seemed possible—the Press" (p. 211), a sure sign that he would find it impossible, as indeed he did. He sees General Grant installed as president, a man who "should have been extinct for ages" (p. 266). At Harvard he discovers history to be a "tangled skein" whose "complexity precedes evolution" (p. 302). Yet the narrator never gives up trying to devise sequences, even though the world seems determined to defeat him. These ruptured sequences suggest that perhaps the most essential part of the dynamo for Adams is not its bipolar configuration or the current that connects the poles but the gap that yawns between them. If the gap were not there, the current could not flow; it is a void as essential to the dynamo's operation as anything explicitly present.

The pattern I want to highlight is a transition from a bipolar to a tripartite construction, with the third part being a rupture or gap. Because a void is normally considered nothing rather than something, transitions to these three-part constructions are often subtle.

However, their characteristically understated emergence should not mislead us into underestimating their importance. The gaps exercise a powerful attraction on Adams's mind, making them not true absences but, as it were, absent presences. To illustrate the process by which a bipolar construction is transformed into a tripartite structure through the insertion of a gap, consider the interplay between the "larger synthesis" Adams took from Hegel (p. 401) and the "larger contradiction" he devised as an answer (p. 406).

For Adams the Hegelian dialectic was closely associated with the scientific relativism he found in Henri Poincaré's *La Science et l'hypothèse* (pp. 454–455). Poincaré pointed out that it was no longer possible, since the advent of non-Euclidean geometry, to ask whether Euclidean geometry was true in any absolute sense. One could ask only if a geometry was consistent, given its axioms. The choice of a geometry was based not on its truth but on its convenience, much as one chooses for convenience a reference point from which other points may be measured. Poincaré saw the development of science in somewhat analogous terms. One accepts at the outset, he argued, that scientific laws are not fixed truths but relative to what is known at the time. If science approaches truth, it does so asymptotically, coming incrementally near but never quite arriving. According to Poincaré, the search for scientific laws proceeds according to a dialectic that alternates between the simple and the complex. When a field of scientific inquiry is new, simple explanations are adequate; as more facts are discovered, explanations grow correspondingly complex. After a sufficient level of complexity is reached, a new synthesis makes things simple again by subsuming previously unassimilated facts into a single explanatory framework. As still more facts are discovered, the tidiness of the theory begins to unravel, until eventually another larger synthesis emerges to "unify the anarchy again" (p. 401).

While admitting the force of Poincaré's argument, Adams takes issue with the implication that a larger synthesis, by virtue of being more encompassing, represents progress. The deciding factor is the human scale. Adams writes that for "human purposes a point must always be soon reached where larger synthesis is suicide" (p. 402). Against the idea of a "larger synthesis" he poses a "larger contradiction," exemplified in the political party he founds with Bay Lodge, the Conservative Christian Anarchists. The party operates according to a dynamic that is dialectical, but with a logic closer to Borges than to Hegel. "By the necessity of their philosophical descent, each member of the fraternity denounced the other as unequal to his loftly task and inadequate to grasp it. Of course, no third member could be so much as considered, since the great principle of contradiction could be expressed only by opposites; and no agreement could be conceived, because anarchy, by definition, must be chaos and collision, as in the kinetic theory of a perfect gas" (p. 406).

Aware that agreement to disagree is still agreement, Adams admits that "this law of contradiction was itself agreement, a restriction of personal liberty inconsistent with freedom; but the 'larger synthesis' admitted a limited agreement provided it were strictly confined to the end of larger contradiction." Within this dialectic, the distinction between order and chaos becomes a personal preference, according to where one chooses to exit from the cycle. For his part, Adams chooses disintegration over synthesis, proclaiming that "in the last synthesis, order and anarchy were one, but that the unity was chaos" (p. 406). Amidst the delight the narrator obviously takes in this paradoxical play, an important modification has taken place almost unnoticed: the bipolar dialectic of Conservative Anarchists has been transformed into the triadic configuration of Conservative Christian Anarchists. Why Christian? What is its function in the dialectic?

Early in *The Education*, the narrator had commented on the lack of religious intensity in his childhood, attributing it to Boston's feeling that it "had solved the universe; or had offered and realized the best solution yet tried. The problem was worked out." Still he lingers over the problem, commenting that of "all the conditions of his youth which afterward puzzled the grown-up man, this disappearance of religion puzzled him most" (p. 34). Whereas in *Mont-Saint-Michel and Chartres* Christianity is a highly charged presence, in *The Education* it is a lack, an absence. Thus the insertion of "Christian" between "Conservative" and "Anarchist" converts the phrase into a three-part structure marked by a void. The presence of the gap allows for the possibility of a transformation beyond the reflexive, paradoxical circling of Conservative Anarchists.

The significance of this gap is revealed when Adams tries and fails to recover Christianity as a contemporary force. Surveying the medieval glass and architecture of France, Adams calls himself the "Virgin's pilgrim," seeing himself as a bridge between the era of faith and the contemporary attitude of religious indifference. This bridge collapses, however, when the violent forces of the twentieth century intervene to rupture the connection between Adams and the Virgin. Visiting the cathedral at Troyes, he sees a "bit of paper stuck in a window"; it is a notice of the assassination of V. K. Pleve in St. Petersburg. "The mad mixture of Russia and the Crusades, of the Hippodrome and the Renaissance, drove him for refuge into the fascinating Church of St. Pantaleon nearby. Martyrs, murderers, Caesars, saints and assassins—half in glass and half in telegram; chaos of time, place, morals, forces and motive—gave him vertigo." He decides that he has to abandon the Virgin. To "what purpose had she existed, if, after nineteen hundred years, the world was bloodier than when she was born? . . . The effort for Unity could not be a partial success; even alternating Unity resolved itself into meaningless motion at last." Then comes the resolve that will shape the remainder of the book. "Every man with self-respect enough to become effective, if only as a machine, has had to account for himself somehow, and to invent a formula of his own for his universe, if the standard formulas failed" (pp. 471–472). The strong sense of forward motion in this passage, of coming to a decision if not an answer, illustrates how the eruption of a gap changes the direction Adams is following.

In *The Education*, a gap is never merely a void. Rather it is a fold that conceals or a tear that reveals. The fold and the tear, although they may seem at odds, are not contradictory in Adams's usage, for when a fold is unfolded, it reveals a tear in the narrative fabric—and through this gap chaos pours. The attraction Adams feels toward the gap is intensely ambivalent. On the one hand, the encounter with chaos can be excruciatingly painful; on the other, it evokes in him an urgent need to make sense of the universe. If it were not for these ruptures, the text might circle endlessly between the forward chronology of the character and the retrospective backward pull of the narrator, much as it circled between the paradoxical antitheses of Conservative Anarchists. The gaps provide the energy necessary to move the narrative forward in more than the superficial senses provided by elapsed time and a sequential turning of pages.

The gaps, partaking of a double nature in the fold/tear, have multiple significations. In some contexts they are associated with the "real" self. I remarked earlier on the complex dynamic of revelation and concealment that characterizes the "real" self, whose presence is constituted through the assertion of its absence. In other contexts, the gap is associated with a woman so desired by the "real" self that the two are inextricable. In still other passages, the gap is an entrance through which chaotic forces pour into the ordered world of the character's certainties.

The chapter called "Chaos" marks Adams's initiation into the agonizing, energizing void for which the gap is the textual signifier. The initiation occurs when he attends his sister Louisa as she lies dying of tetanus. His horror at her suffering is intensified by the contrast between her racking pain and the opulent beauty of the Italian summer outside the sickroom. "Death took features altogether new to him, in these rich and sensuous surroundings. Nature enjoyed it, played with it, the horror added to her charm, she liked the torture, and smothered her victim with caresses." The shock strips from his perception the ordered forms that normally populate it, plunging him into a vision of the universe that he will later associate with the random motion of gas molecules in the kinetic theory of heat. The "stage-scenery of the senses collapsed; the human mind felt itself stripped naked, vibrating in a void of shapeless energies, with resistless mass, colliding, crushing, wasting, and destroying what these same energies had created and labored from eternity to perfect." The normal routine of society is revealed as a "pantomime with a mechanical motion," its "so-called thought merged in the mere sense of life, and pleasure in the sense" (p. 288).

Traveling to the Alps to recover his balance, he finds that for "the first time in his life, Mont Blanc for a moment looked to him what it was—a chaos of anarchic and purposeless forces—and he needed days of repose to see it clothe itself again with the illusions of his senses" (p. 289). At this moment the character's perception merges with the world of the "real" self, for when he senses that he has been surrounded by a kind of stage scenery, he comes close to intuiting his own status as a manikin. The effect is not so much to emphasize the character's artificiality, however, as to reveal the constructed nature of consensual reality, for the theater of stage scenery turns out to be the world we all share.

Many readers have seen in Adams's description of his sister's agony the death he does not describe—the suicide of his wife, Marian, who took her own life at age forty-two by swallowing potassium cyanide from her darkroom.[7] Very probably the bonds that join the "real" self to the woman and to chaos were forged in the crucible of Adams's suffering after his wife's death. Marian's death falls into the twenty-year silence that stretches between the first and second halves of *The Education*. If we think of this lacuna not only as a bottomless hole into which things disappear but also as a rent from which a resistless, remorseless energy radiates, we can better understand the force that animates the character Adams after he declares himself "dead" following his wife's death (p. 330).

Once the gap opens and chaos begins to pour through, it is only a matter of time until chaos becomes a flooding tide on which the remaining pockets of order float like so many isolated islands. To see how this transformation is embodied in the text's language, compare two passages, one from the opening pages, the other near the end. In the first, in a major statement of his theme, the narrator declares that the "problem of running order through chaos, direction through space, unity through multiplicity, has always been, and must always be, the task of education, as it is the moral of religion, philosophy, science, art, politics, and economy" (p. 12). The lines of sight in this passage run horizontally, an orientation reinforced by the series of parallel phrases. The active principle is order; words associated with chaos are imaged as mere inert media through which direction and unity run. The activity of "running" is, moreover, itself a unifying principle, for it joins in a common endeavor the arts and sciences, as well as the work performed by education (and by implication, *The Education*).

Contrast these characteristics with a much later passage in which Adams, musing on the new psychology, thinks that it conceives of the mind as a bicycle rider "mechanically balancing himself by inhibiting all of his inferior personalities, and sure to fall into the sub-

[7]See, for example, Rowe, 1976:116; and Paul Hamill's "Living and Dying in History: Death in *The Education of Henry Adams*," *Soundings* 60 (1976): 150–65, especially p. 153. For a speculative treatment of the effect on Adams's writing of his wife's death, see Elizabeth Waterston, "The Gap in Henry Adams's Education," *Canadian Review of American Studies* 7 (1976): 132–138. The fullest source remains Ernest Samuels, *Henry Adams: The Major Phase* (Cambridge: Harvard University Press, 1964).

conscious chaos below, if one of his inferior personalities got on top. The only absolute truth was the sub-conscious chaos below, which everyone could feel when he sought it" (p. 433). Here the lines of sight are vertical, plummeting down into chaos or rising to balance precariously over it. The imagery emphasizes the disjunctions that divide what was a unity into discrete units, all fighting against one another in a cacophony of competing claims—the rider struggling with the bicycle, the bicycle teetering over chaos, the rider divided within himself by warring personalities held only temporarily in check. Instead of parallel phrases, with their suggestion of a broad, stable base, the rhetoric relies on a conjunction that depends on a conditional clause, which is further qualified by the alarming assertion that the "only absolute truth was the sub-conscious chaos below." Although the rider is active to the point of frenzy, the dominant principle, and the force against which he must struggle, is chaos—without and within.

Granted, in this passage Adams is writing about what the new psychology thinks the mind is rather than what he thinks it is. But exposure to chaos in the language of the text does not leave the character Adams unchanged, any more than exposure to chaos in the world has left humankind unchanged. Increasingly as *The Education* progresses, the character Adams finds himself drawn, through the narrator's rhetoric, into the "void of shapeless energies" that he intuited behind the stage scenery of ordered representations when Louisa died.

The phrase the narrator uses to signify chaos flooding into his text (and into the world) is "supersensual chaos" (p. 451). The phrase derives from Karl Pearson's *Grammar of Science*, summarized in the chapter of that name in *The Education*. According to Pearson, scientific representations are necessarily limited to the realm of sensory impressions. Of the "chaos behind sensations," science can know nothing. In a passage that Adams quotes, Pearson writes " 'in the "beyond" of sense-impressions, we cannot infer necessity, order or routine, for these are concepts formed by the mind of man on this side of sense-impressions.' " The narrator's contribution to Pearson's image is to adjust the scale of proportion, so that the ordered world of human perception appears as a speck floating on an ocean of chaos. The change of proportion is evident in his paraphrase of Pearson's assertion: "Chaos was the law of Nature; Order was the dream of man" (p. 451). The sentence exemplifies the paradoxical play that unites chaos with order, absence with presence in *The Education.* Read straightforwardly, it asserts that chaos has displaced order as the texture of reality; but in the process chaos has become a law, indicating that chaos itself is a kind of order. Order, for its part, is identified with dream, signifying its unreality. Since the "dream of man" is opposed to the "law of Nature," however, there is also the suggestion that order is the reality we know, whatever the case may be for reality-in-itself.

These implications are made explicit in a longer passage that sums up woman's sense of order, in images that extend and ramify the connotations of a supersensual chaos. "She did not think of her universe as a raft to which the limpets stuck for life in the surge of a supersensual chaos; she conceived herself and her family as the centre and flower of an ordered universe which she knew to be unity because she had made it after the image of her own fecundity; and this creation of hers was surrounded by beauties and perfections which she knew to be real because she herself had imagined them" (p. 459). An unreal order that materializes into presence because it has been imagined; and a chaos that fades into absence because it exists in external reality. The dialectic is the same as that used to constitute the multiple personae of Henry Adams: the character who materializes into a textual presence because he has been imagined; the "real" self who fades into absence because he exists in external reality. It is thus not only through images that chaos and the "real" self are associated. They are also conjoined because they occupy the same structural positions within the dynamics of the text.[8]

[8]The structural connection that *The Education* posits between the self and the void has a parallel in the psychoanalytic theories of Jacques Lacan, especially in Lacan's assertion that it is the *absence* of the object that leads a child to desire it, or indeed to recognize it as an object. For Lacan, it "is the world of words which creates the world of things" (*Speech and Language in Psychoanalysis*, trans. Anthony Wilden [Baltimore: Johns Hopkins University Press, 1981], p. 39). Thus it is only with a child's entrance into language that the world of objects comes into existence. Since the essential psychical dynamics derive from absence rather than presence, it follows that the world of objects is also constituted as a series of successive absences. Presence arrives through metonymic displacement of one absent object by another, just as signification is built upon a chain of signifiers that emerge from a void. Where Adams departs from Lacan is precisely the point that makes *The Education* of interest for this study—in associating the void with chaos. While recognizing that the parallel with Lacan exists, I prefer to develop my argument about *The Education* through a vocabulary and a frame of reference closer to Adams's own thinking.

As Adams the character gets drawn deeper into the gap that is at once chaos and his own mind, he becomes more determined to find a sequence that can connect past to future, self to world, order to chaos. The importance of sequence in establishing a connection between self and the world goes back at least to Descartes. In his 1637 *Discourse on Method*, Descartes drew a parallel between sequences in Euclidean geometry and the inductive chains of reasoning that allowed him to connect the thinking mind with the outside world. He implied that the self can know the world *because* reliable sequences can be constructed. "The long chains of simple and easy reasoning by which geometers are accustomed to reach the conclusions of their most difficult demonstrations, led me to imagine that all things, to the knowledge of which man is competent, are mutually connected in the same way, and there is nothing so far removed from us as to be beyond our reach or so hidden that we cannot discover it, provided only we abstain from accepting the false for the true, and always preserve in our thoughts the order necessary for the deduction of one truth from another."[9] Reading Descartes, one can better appreciate Adams's shock in learning from Poincaré that Euclidean geometry is not true, merely convenient. If geometry is the standard against which the construction of sequence is to be measured, what does it signify that the chains of reasoning Descartes so admired are merely extrapolations from a point chosen because it was convenient?

The task Adams set himself was to wrest from the void a new conception of sequence that would inhere in the very forces that disrupted traditional sequences. Finding that the "sequence of men led to nothing," that the "sequence of time was artificial," and (in a revealing phrase) that the "sequence of thought was chaos," Adams "turned at last to the sequence of force" (p. 382). To make the gap give birth to this new kind of sequence, he refashioned it into an anarchical space crackling with violent energies—an image brought to life by the dynamo. The result was the dynamic theory of history, presented in *The Education* as Adams's own attempt at a larger synthesis. The theory rewrites the history of humankind as its encounter

[9]Quoted in Morris Kline, *Mathematics: The Loss of Certainty* (New York: Oxford University Press, 1980), p. 98.

with increasingly powerful forces. Sequences are disrupted when a new and more anarchical force enters the stage of human action. Forces can be either physical or spiritual; Christianity qualifies as an epoch-initiating force, along with gunpowder and fire. The theory predicts that the world will end with a bang rather than a whimper, for the process will continue at an accelerating rate until the power of the captured forces exceeds the ability of humankind to control them.

Throughout Adams's account of his dynamic theory, as John Carlos Rowe has noticed, there is a disturbing ambiguity about the direction in which the lines of force run.[10] We are told, for example, that the dynamic theory "takes for granted that the forces of nature capture man." But within a few lines we are presented with an image that shows man capturing the forces. The theory "may liken man to a spider in its web, watching for chance prey. Forces of nature dance like flies before the net, and the spider pounces on them when it can." Who is in control, man or the forces? "The sum of force attracts; the feeble atom or molecule called man is attracted; he is the sum of the forces that attract him; he suffers education or growth; his body and his thought are alike their product; the movement of the forces controls the progress of his mind, since he can know nothing but the motions which impinge on his senses, whose sum makes education" (p. 474). Adams seems to imply that man only thinks he is capturing forces, whereas in reality they are capturing him; but after the mind and chaos have become sufficiently intertwined, it is difficult to tell which has priority. For if man has become "the sum of the forces that attract him," is he not transfigured into that which shaped him, thus becoming his own origin?

"The universe that had formed him took shape in his mind as a reflection of his own unity, containing all forces except himself" (p. 475). Is this a description of the way the universe interacts with humans? Or of *The Education of Henry Adams*, whose encyclopedic scope seems to include everything but the "real" Henry Adams? How can we distinguish between a mind formed by a universe of

[10]Rowe comments that the "image of the spider and its prey is inverted curiously. In a conventional deterministic scheme, *man* would be the prey of forces beyond his control. Yet here the forces are the victims of the trap, the mind acting as the weaver of the pattern" (1976:127).

chaotic forces and a mind that is the origin for those same chaotic forces? Consider: the proof that man has been formed by a universe of force is a turning inside out of the human mind, in which it becomes a container for the forces but no longer contains itself. Where in this construction is the self located? Presumably exterior to the mind that contains "all forces except himself." Thus the void becomes the self and the self becomes the void, each occupying the other's former position as if they were the inside and outside surfaces of a Möbius strip.[11]

Implicit in these inversions is a realization central to the pattern I have been describing. *The Education of Henry Adams itself embodies the process it describes in the dynamic theory of history*. The sequence of the text is established not by Adams's personal history or by the history of his time but by the continual interaction of his mind with the chaotic forces of the void.[12] Although he resists, at each encounter Adams absorbs more of the chaos within himself, until his mind is sufficiently energized to give form to the chaos without, as he does by postulating the dynamic theory of history. "Clearly if he was bound to reduce all these forces to a common value, this common value could have no measure but that of their attraction on his own mind. He must treat them as they had been felt; as convertible, reversible, interchangeable attractions on thought" (p. 383). The inside/outside inversion that hovers just on the edge of articulation in this passage embodies a profound ambiguity about where the chaos is finally to be located. Once Adams has absorbed enough of it, it becomes he and he becomes it. Ronald Martin, in his fine study *American Literature and the Universe of Force* (1981), edges toward this realization when he points out that the "Dynamic Theory and the process of its formulation are con-

[11]Jacques Lacan discusses how the inside/outside inversion relates to his construction of self on the basis of absence in "Of Structure as an Inmixing of an Otherness Prerequisite to Any Subject Whatsoever," In *The Languages of Criticism and the Sciences of Man: The Structuralist Controversy*, ed. Richard Macksey and Eugenio Donato (Baltimore: Johns Hopkins University Press, 1970), pp. 186–200, especially 192.

[12]G. Thomas Couser, in "The Shape of Death in American Autobiography," *Hudson Review* 31 (1978): 53–66, comments that "upon close examination of the theory, we find that in spite of his habitual self-deprecation, Adams put his own consciousness squarely at the center of his own cosmology. . . . His formula for the universe and his account of himself were aspects of the same vision" (p. 63).

foundingly metaphorical: the protagonist's career serves as a metaphor for the course of Western civilization, but finally the course of Western civilization serves as a metaphor for the narrator's mind. There is no simple way to take it" (p. 139).

The self-reflexive circling that connects the self to the chaotic void has a moral aspect that is capable of invoking almost palpable pain in the narrator. If Adams has become the chaos he deplores, how can he fix the limits of his responsibility? Contemplating the accelerating forces to which he believes human history is subject, the narrator confesses that "all that a historian won was a vehement wish to escape. He saw his education complete. . . . He repudiated all share in the world as it was to be, and yet *he could not detect the point where his responsibility began or ended*" (p. 458; emphasis added). But perhaps Adams, rather than bringing chaos into the world because his mind attracts it, is merely the instrument of forces that attract his mind. Some of the images suggest this more passive possibility. For example, the narrator, in a continuation of the passage cited earlier on the new psychology, proposes that the new science of the mind sees it as an electromagnet, "mechanically dispersing its lines of force when it went to sleep, and mechanically orienting them when it woke up." He impishly wonders "which was normal, the dispersion or orientation?" (p. 434). The prospect that one's mind is an electromagnet automatically generating lines of force instead of the lines of will Adams resolved to follow in "Darwinism" may not be flattering, but it at least alleviates the sense of guilt.

A more disturbing idea is the inference Adams draws from his dynamic theory, that the "laws of history only repeat the lines of force or thought" (p. 457). If this is true, then Adams's thoughts are partly responsible for bringing into existence a world so chaotic that, by the end, the narrator must wonder whether "sensitive and timid natures could regard [it] without a shudder" (p. 505). The narrator tries hard to maintain his objectivity. Earlier he laid down as an article of faith that no "honest historian can take part with—or against—the forces he has to study" (p. 447). But faced with his responsibility for bringing chaos into the world, as his own dynamic theory implies, the narrator admits that "though his will be iron, he cannot help now and then resuming his humanity or simianity in face of a fear." What is this fear? "The motion of thought had the same value as the motion of a cannonball seen approaching the observer on a direct line through the air. One could watch its curve for five thousand years. Its first violent acceleration in historical times had ended in the catastrophe of 310. The next swerve of direction occurred toward 1500. Galileo and Bacon gave a still newer curve to it, which altered its values; but all these changes had never altered the continuity. Only in 1900, the continuity snapped" (p. 457). Toward whom is the cannonball heading now, and what force is directing its path? The narrator's worst nightmare is that the answer to both questions is the same: the mind of Henry Adams. In this vision of terror, Adams cannot escape the cannonball because he cannot elude his own thoughts. More literally, he cannot escape because he *is* the cannonball.[13]

This nightmare vision explains why establishing the proper sequence is the central problem of *The Education*. If Adams, like the spider, pounced on the forces, providing first through his consciousness and then through his text the gaps through which it could flood the world, then he must accept responsibility for creating a multiverse he deplores. If, on the contrary, the forces captured him, then he is at most a helpless and passive accomplice to the inevitable. The problem reduces to a question of sequence: did the textual gaps create the chaos that emanates from them, or did the resistless force of chaos create the gaps when it tore the fabric of order and textuality? Everything is made to hang on the answer, from the fate of the world to the moral responsibility Adams must accept in helping to create that fate.

It is not difficult to read into this urgent questioning a hidden agenda for also establishing the extent of his responsibility for the fate of the woman. Pondering on why woman has ceased to be the potent force that Adams believes is her birthright by virtue of her sexual power, the narrator suggests that "at dinner, one might wait till talk flagged, and then, as mildly as possible, ask one's liveliest neighbor whether she could explain why the American woman was a failure. Without an instant's hesitation, she was sure to answer:

[13]I am indebted to Sandra Bernhart, in a paper written for my Adams and Pynchon Seminar at Dartmouth College, 1981, for the suggestion that the cannonball in this passage is Adams's thought. It uncannily looks forward to Pynchon's use of Enzian as a human cannonball in the Rocket. Perhaps the correspondence is not accidental. Pynchon was deeply influenced by *The Education of Henry Adams*, as both *V.* and *Gravity's Rainbow* reveal.

'Because the American man is a failure!' She meant it" (p. 442). Whether this remark had a deeper personal significance for Adams in view of the tragedy that marked his private life we cannot know, although it is a likely surmise. Within the context of *The Education*, the woman who most obviously fails because he (and everyone else) has failed her is the Virgin. Like most of the failures in *The Education*, the Virgin's has an ambiguous value. Although she has ceased to be a force in contemporary society, she nevertheless serves as a reference point from which the force of the Dynamo may be measured. Thus even in her absence, she serves as an anchor for Adams's new kind of sequence. By implication, the woman also serves as a reference point from which the self may measure its responsibility for creating the chaos it abhors.

To make his new sequence work, Adams needs to establish that the units used to measure the two different kinds of forces represented by the Dynamo and the Virgin are equivalent. This means, of course, that he has to find a common measure for mechanical and spiritual (i.e., sexual) attraction. That measure is his mind. "He must treat them as they had been felt; as convertible, reversible, interchangeable attractions on thought. He made up his mind to venture it; he would risk translating rays into faith" (p. 383). The subjectivity that informs Adams's use of metaphor is not merely whimsical, any more than the form of *The Education* is arbitrary. On the contrary, it is a rigorous working out of the conditions that must obtain if Adams is to create a nontrivial sequence that will connect him to a chaotic multiverse and also allow him to confine chaos within the laws of order. The connection between the "real" self, the woman, and the chaotic void is a knot woven into the very texture of this text. If it were untied, the logic that holds the text together would also unravel.

Through this knot, Adams arrives at a transvaluation of chaos not unlike that the scientific tradition achieved in the century between 1850 and 1950. Although Adams followed a unique route to arrive at his result, there are significant parallels between *The Education* and the scientific texts. Especially striking is the key role that metaphors play in both transformations. In the scientific texts, metaphors embody connotations not explicitly present in the theories but nevertheless instrumental in guiding their development. In *The Education*, the connection between metaphors and the underlying thought lies much closer to the surface. Metaphors in *The Education* characteristically act as juncture points between subjective experience and objective reality. Consider, for example, the "Darwinism" chapter as the site of a suture. This chapter introduces two biological terms—*Terebratula* and *Pteraspis*—that reappear throughout the text, each time with increasing resonance. Tracking the evolution of these terms will illustrate how *The Education* compounds the subjectivity of its personae with the apparent objectivity of scientific concepts.

We meet the first term, *Terebratula* (designating a genus of brachiopods), when Sir Charles Lyell tells Adams it is a life form that "appeared to be identical from the beginning to the end of geological time" (p. 228).[14] Although Sir Charles seems oblivious of the implication that *Terebratula* contradicts the continuous, unbroken evolution that he champions, to Adams it indicates "altogether too much uniformity and much too little selection" (p. 228). A closely related term is *Pteraspis*, designating a ganoid fish and "cousin to the sturgeon." *Pteraspis*, the first vertebrate to appear in the fossil record, resided in a "kingdom . . . called Siluria" (p. 229). The narrator calls it "one's earliest ancestor and nearest relative" (p. 228). Like *Terebratula*, *Pteraspis* also poses a challenge to Darwinism, for beyond it stretches only a void rather than the graduated development that continous evolution predicts.

Terebratula and *Pteraspis* reflect the double vision of Adams presented in *The Education*.[15] On the one hand the narrator characterizes the young Adams as an antiquated mollusk like the *Terebratula*, himself unchanging while the world breaks from its moorings around him. On the other hand, Adams the character comes increasingly to suspect that hidden within him is another, catastrophic ver-

[14]Edward Morse, a British scientist who traveled to Japan in the 1850s, proved that the *Terebratula* was in fact not a mussel but a worm. An account of Morse's discovery can be found in Robert A. Rosenstone, *Mirror in the Shrine: American Encounters with Meiji Japan* (Cambridge: Harvard University Press, 1988). Allusions in *The Education* suggest that Adams was unaware of the distinction. I am grateful to Professor Rosenstone for showing me pictures of *Terebratula*, which look like tiny mussels a fraction of an inch in diameter.

[15]In my analysis of these metaphors, I have drawn on R. P. Blackmur's discussion of them in the chapter called "*Terebratula* and *Pteraspis*" in his *Henry Adams* (1980:62–72). Wayne Lesser also has an illuminating discussion of them in "Criticism, Literary History, and the Paradigm: *The Education of Henry Adams*," *PMLA* 97 (1982): 378–94.

sion of himself which will emerge like the *Pteraspis*, rupturing sequence and making continuity with the past impossible. It is no accident that this realization occurs in the chapter where the terms are first introduced. The contradiction between stasis and rupture plays through subsequent uses of the terms, making the supposedly objective reality they represent as convoluted and turbulent as Adams's own subjectivity.

Consider, for example, those instances where the narrator uses the terms metonymically to mark places where sequences should have begun but did not. Despairing of using time as a meaningful sequence, the narrator remarks that he "could detect no more evolution in life since the *Pteraspis* than he could detect it in architecture since the [Wenlock] Abbey. All he could prove was change" (p. 230). The lack of correlation between time passed and evolution achieved applies also to him. After being in England eight years, he found that he "was already old in society, and belonged to the Silurian horizon" (p. 235). The implication that he belongs to an age discontinuous with the present is confirmed when he arrives in Washington and discovers that the connections his family afforded him are merely the "accumulated capital of a Silurian age. A few months or years more, and they were gone" (p. 252). But the arrival of the new postwar administration with the election of General Grant is so regressive that it makes him look progressive by comparison. "Grant fretted and irritated him, like the *Terebratula*, as a defiance of first principles. He had no right to exist. He should have been extinct for ages" (p. 266). Embodied in these references is an increasing ambiguity about whether Adams is atavistic or chaotic, moored to the past or already part of a future discontinuous with the present.[16]

[16]With this nimbus of meaning surrounding the terms, their presence anticipates as well as confirms the confounding of past and future. When Adams is asked to teach history at Harvard, he remarks that "even a *Terebratula* would be pleased and grateful for a compliment which implied that the new President of Harvard College wanted his help" (p. 291). The implication is that an unchanging Adams will not be able to teach history effectively, because he cannot establish anything other than the most artificial of sequences. He finds that "nothing is easier than to teach historical method, but, when learned, it has little use. History is a tangled skein that one may take up at any point, and break when one has unravelled enough; but complexity precedes evolution." He is also kept from a view of history as stasis, however, because the "*Pteraspis* grins horribly from the closed entrance" (p. 302).

Another cluster of references comes in the "Twilight" chapter when Adams, convinced that the confounding of sequence is pervasive in his own life, looks to science to verify that ruptured sequences are objectively true as well. He returns to biology to discover "what had happened to his oldest friend and cousin the ganoid fish, the *Pteraspis* of Ludlow and Wenlock, with whom he had sported when geological life was young; as though they had all remained together in time to act the Mask of Comus at Ludlow Castle, and repeat 'how charming is divine philosophy!'" (p. 399). Predictably, he finds that, although an ancestry has been discovered for *Pteraspis* in the Colorado limestone, the field of geology is in a state of confusion that surpasses his own. "The textbooks refused even to discuss theories, frankly throwing up their hands and avowing that progress depended on studying each rock as a law to itself" (p. 401). The narrator is not optimistic that some new theory will arrive to make sense of the contradictions. All that he can be sure of is "rapidly increasing complexity," although he wonders if "the change might be only in himself" (p. 402).

The compounding of past and future, stasis and rupture, becomes explicit in the brilliant artifice of the paradoxes that glitter at the end of the "Twilight" chapter. Intuiting that he is himself an instigator of the change he loathes, the narrator longs to "begin afresh" with ancient life forms in prehistoric waters, "side by side with Adamses and Quincys and Harvard College, all unchanged and unchangeable since archaic time." But then he realizes that if change has indeed originated within him, stasis is doomed in any event, even if he could return to the beginning. Somehow the rupture was forecast at the beginning, even as the stasis of the beginning lingers as a longing for return after the rupture has occurred. He concludes that a "seeker of truth—or illusion—would be none the less restless, though a shark!" (p. 402). The sentiment serves as conclusion for the chapter and marks a moment in the text when no sequence connecting past and future seems possible, except the sequence established by the repeated and pervasive failure of all previous sequences.

The patterns formed by this evolving network of references demonstrate how profoundly Adams merges technical denotation with the deeper thematic and psychological dynamics of his text. Once chaos and contradiction are recognized as important constituents of subjectivity, it is imperative to Adams that they be present in objective reality as well. Only thus can his consciousness be joined to the multiverse that he believes represents the world as it is or will be. To admit the possibility that science could demonstrate that the multiverse was in fact a universe, or that objective reality was ordered, would in the face of a chaotic subjectivity be an invitation to solipsism, if not to unrelievable guilt.

William Jordy, in his definitive study *Henry Adams: Scientific Historian* (1952), shows how Adams systematically distorted and misrepresented the scientific evidence to make sure that the correlation between his chaotic subjectivity and a chaotic world would obtain. But Adams was premature in his judgment that science was in disarray. Even as he was writing *The Education*, one of the most successful of science's larger syntheses had already been published—Einstein's special theory of relativity. It would be another quarter of a century before the new synthesis began to unravel again with quantum mechanics, and a half century before the chaos that so preoccupied Adams would become important in scientific theorizing.

Perhaps the most puzzling of Adams's scientific (mis)appropriations is his vision of chaos as an energizing force capable of stimulating self-organization and increasing complexity. Granted, complexity did not have for Adams the entirely positive connotations it has in contemporary scientific discourse. Nevertheless, in seeing that chaos could be linked to the emergence of complex systems, Adams did envision chaos very differently than had Kelvin or even Boltzmann. Yet a mere three years after he had completed *The Education*, Adams published "A Letter to American Teachers of History" (1910), in which he used classical thermodynamic arguments to demonstrate that the world and human intellect were running down. The sophisticated reasoning characteristic of the modern mind represented an entropic decline; the high point for thc human race had come in the Middle Ages, before instinct had wholly given way to reason.[17] How could he espouse this entropic view of history when

[17]John J. Conder, in *A Formula of His Own: Henry Adams's Literary Experiment* (Chicago: University of Chicago Press, 1970), argues that the decline from instinct to reason forms the structural principle for all of Adams's late works, including *The Education*. To my mind, this orientation causes Conder to misread some parts of

he so recently had worked out the implications of an accelerating view in *The Education*?

As Jordy points out, it is possible to reconcile the entropic and accelerating views if one assumes that intellectual speculation and physical energy can be equated, and that both operate within a closed system (1952: 158–219). If humans are thinking in ever more complex ways, they are presumably depleting their store of intellectual energy at increasing rates, and thus declining faster and faster even while they appear to be accelerating. Of course, the premise that human evolution takes place within a closed system is incorrect. Life is possible only because our planet receives massive doses of energy from the sun every day. Similarly, the assumption that intellectual and physical energy can be equated is at the very least extremely problematic, as Professor Henry A. Bumstead of Yale informed Adams when he was asked to respond to a manuscript draft of "The Rule of Phase Applied to History."

After receiving Bumstead's criticism in January 1910, Adams made only small revisions in "A Letter to American Teachers of History" before mailing it out to college presidents, professors, and libraries in the spring of 1910. Later that year he revised "A Rule of Phase Applied to History" (originally published in 1909), taking into account some of Bumstead's corrections but allowing the central thesis to stand.[18] It is possible that Adams simply did not understand the gravity of Bumstead's objection that thought and energy cannot be equated. So customary—and necessary—was his compounding of human consciousness with the supposedly objective world of chaotic force that for him to take full account of it would have required a complete reorientation of his thought—a task that, at age seventy-two, he may well have been unable as well as unwilling to undertake.

The Education where infolded complexity is most apparent, such as the "Darwinism" chapter. Nevertheless, his study is valuable in highlighting those aspects of *The Education* that forecast most clearly the entropic view of the later works.

[18]Jordy (1952) has a full discussion of Bumstead's criticism and the chronology of revisions (pp. 152–153, 273ff). Both "A Letter to American Teachers of History" and "The Rule of Phase Applied to History" were published by Brooks Adams, with an introduction, in *The Degradation of Democratic Dogma* (1919).

Ronald Martin, reviewing the evidence of the manuscript revisions and the Bumstead comments, concludes that Adams's "grasp of modern science does seem to have been as weak as he claimed it was."[19] Martin argues that the contradiction between the self-organizing view of chaos in *The Education* and the entropic view of it in the later scientific essays, although so obvious that it could scarcely be missed, was of little or no consequence to Adams. "It did not seem to bother him that instead of the universe accelerating itself to its death in an explosion of etherealization . . . it now promised to degrade itself into a dead ocean of entropy. What mattered was that the end be reputably scientific and that it be soon" (1981:143).

Martin may well be correct as far as Henry Adams, the historical figure, is concerned. However, for the text of *The Education* and the multiple personae called Henry Adams constituted within it, it makes all the difference that chaos is conceived as capable of creation as well as destruction. It is this ambiguity that authorizes the narrator to dream he might be drawn into the gap and find there an equi/multivocal unity amidst his multiple selves and the world. "There is nothing unscientific in the idea that, beyond the lines of force felt by the senses, the universe may be—as it has always been—either a supersensous chaos or a divine unity, which irresistibly attracts, and is either life or death to penetrate" (p. 487). It is this ambiguity that allows him to imagine the child of the twentieth century as "a sort of God compared with any former creation of nature" (p. 497). And finally, it is this that encourages him to imagine that he and his two closest friends, united in 1938 on the centenary of their births, "might be allowed to return together for a holiday, to see the mistakes of their own lives made clear in the light of the mistakes of their successors; and perhaps then, for the first time since man began his education among the carnivores, they would find a world that sensitive and timid natures could regard without a

[19]To illustrate Adams's scientific ineptitude, Martin points out that he "needed to be told such things as that the terms 'movement,' 'velocity,' and 'acceleration' are not roughly synonymous; that potential energy is not concealed kinetic energy; that some connections Adams calls 'logical' are really analogical; that light is not polarized by a magnet; that entropy is not merely deadness; that 'lines of force' are not a physical reality; that Adams applied what he called 'the law of inverse squares' incorrectly, and that he even neglected to figure it inversely" (1981:142).

shudder" (p. 505). However qualified and ironic this hope for a better world is, it is an improvement over the narrator's earlier vision of himself, alone, facing the plummeting cannonball that is also his thought.

In light of Adams's scientific inaccuracies, it would be misleading to picture him as a prophetic thinker who anticipated the science of chaos when scientists themselves were only beginning to understand the limitations of the Newtonian paradigm.[20] Rather, his importance for this study lies in the subtle and complex connections he established between chaos and the way the self is constituted within language and literature. In associating the self with absence rather than presence, he negotiated some of the same territory that would later be staked out by such seminal theorists as Lacan and Derrida. In connecting this problematic self to chaos, he devised strategies of representation that foreshadowed the work of such important contemporary writers as Stanislaw Lem, Doris Lessing, and Italo Calvino. One of the punning connections that Adams was fond of making in *The Education* was between his name and Adam, first man on earth. The pun expressed his rootedness in the past and, increasingly as *The Education* progressed, the fear that he himself was the catastrophe that severed past from future. From our perspective, the chaos that tore him from his roots is also the thread connecting him to the future he dreaded and anticipated. It is an irony he would have appreciated.

[20]Paul Hamill, in "Science as Ideology: The Case of the Amateur, Henry Adams," *Canadian Review of American Studies* 12 (1981): 21–35, defends Adams's use of science by arguing that the "scientific ideas that engaged him were those that still possessed an unexplored metaphorical dimension for scientists too. No one was entirely sure how far a strictly scientific understanding of evolution or entropy could be extended to explain human life and culture" (p. 23). I concur; but this is different from saying that Adams foresaw the scientific theories. Ernest Samuels, in his summary article "Henry Adams: 1838–1918," in Harbert, 1981:84–103, comments that although we cannot take the science of *The Education* seriously, Adams shows "virtuosity" in manipulating the new concepts, demonstrating that "a new field of metaphor is available to the literary artist and that scientific reality can add a new idiom to poetry" (pp. 102–133).

CHAPTER 4

From Epilogue to Prologue: Chaos and the Arrow of Time

ILYA PRIGOGINE and Isabelle Stengers's *Order out of Chaos* (1984) should carry a warning label: "CAUTION: Use at own risk. Authors have speculated beyond data. Conclusions are conjectural." The disclaimer is necessary because this book is not, as it may appear to be, simply a popularized version of the new sciences. Rather, it is an ambitious synthesis that goes well beyond what many scientists working in chaos theory would be willing to grant are legitimate inferences from their work. As speculative theory, it attempts to define the deeper significance—one might almost say the metaphysics—of an emerging area of research. At issue is whether chaos should be associated with the breakdown of systems or with their birth. Viewed as the epilogue to life, chaos almost inevitably has negative connotations. Seen as prologue, it takes on more positive evaluations.

At the center of Prigogine and Stengers's project is their view of how the new science relates to time. According to them, chaos theory provides a resolution to the long-standing philosophical debate about whether being or becoming is the essential reality. They point out that classical physics imagines the universe as reversible, able to move either backward or forward in time. The biological and human sciences, by contrast, locate themselves within an irreversible world of birth and mortality. Prigogine and Stengers see thermodynamics as forming a natural bridge between these views. At equilibrium, thermodynamic equations are classical in the sense that they are reversible, but far from equilibrium they are irreversible and hence bound to a one-way direction in time. A thermodynamics that integrates reversibility with irreversibility would, they assert, allow being and becoming finally to be reconciled.

Even from this brief summary, the reader can appreciate the disparity between the sweep of the theoretical claims and the technical basis on which they rest. To imagine that profound philosophical questions can or should be answered by advances in irreversible thermodynamics is apt to evoke uneasiness among most scientists and outright skepticism among humanities scholars. Yet there have been scientific theories that have significantly affected social attitudes—evolution, relativity, and quantum mechanics, for example. Whether chaos theory will have this kind of impact remains to be seen. Even more problematic, in my view, is whether the new paradigms are causes of social change or are themselves reflections of larger cultural currents. In light of these uncertainties, it is especially important to be clear about what different research programs within the sciences of chaos have accomplished, and what they have so far merely promised or suggested.

The even tone in which *Order Out of Chaos* is written does not encourage this distinction. Facts that have been extensively verified are presented in the same manner as speculations backed by little or no experimental confirmation. Only once do Stengers and Prigogine acknowledge that some of their scientific claims are conjectural. At the end of the chapter on microscopic interpretations of irreversibility, they add a "word of caution" that "experiments are in preparation to test these views," and as "long as they have not been performed, a speculative element is unavoidable" (p. 290). The implication, of course, is that the "speculative element" waits only on the completion of experiments to be removed. Since no attempt is made in the chapter to identify why, or even what, points are speculative, the lay reader has no way to evaluate the claim that experimental confirmation is forthcoming. The "word of caution" is just that—a brief comment added as an afterthought, not an acknowledgment that many of the chapter's conclusions are controversial.

Additional difficulties are presented by an inference many readers have drawn from *Order out of Chaos*, that Prigogine is as central to chaos theory as Darwin was to evolution or Einstein to relativity. To someone who works in the field, it appears that Prigogine's work has supplied a few rooms in a large building. Certainly he did not establish the building's architecture, nor did he single-handedly lay the foundation. To be fair, the misapprehension is not entirely his fault. It is perhaps understandable that in a text he co-authored, his research would be foregrounded and other research programs reported as though they merely provided corroborating evidence for his thesis. Nevertheless, the imbalance implicit in *Order out of Chaos* should be pointed out and corrected, not replicated.

Despite these problems, *Order out of Chaos* is an important work. It is frequently cited by scholars interested in connections between literature and science because it is one of the very few texts available that attempt to explain the new sciences in words rather than equations. Moreover, the research it reports is interesting and worthwhile, independent of its philosophical significance. But as we shall see, its most important claims have to do with changes in world view rather than shifts in scientific theorizing. To see how these claims emerge, it will be necessary to review Prigogine's contributions to irreversible thermodynamics. For that we need to return to the second law of thermodynamics.

As we saw in chapter 2, the second law led to predictions of a "heat death" in which the universe, having exhausted all heat reservoirs, degenerates to a state of maximum dissipation. This scenario contrasts sharply with the spontaneous increase of organization in living organisms, as in the increasing differentiation of an embryo. Twenty years after Kelvin formulated the idea of "universal dissipation," he agonized in print over whether living organisms were an exception to it. He finally concluded it was futile to worry about the matter. "But the real phenomena of life infinitely transcend human science," he wrote, "and speculating regarding consequences of their imagined reversal is utterly unprofitable" (1911: 12). The apparent discrepancy between entropic decay and biological complexity can be reconciled when one recognizes that the second law applies only to closed systems, whereas all living organisms are open systems that continually receive energy from outside themselves. Nevertheless, the contrast was troubling to Kelvin, and remained so for many thermodynamicists. Biologists, for their part, tended to feel that thermodynamics was largely irrelevant to their concerns.

Prigogine gave a new twist to this old problem by reformulating the way entropy production takes place. He reasoned that the overall entropy term can be divided into two parts. The first reflects exchanges between the system and the outside world; the second describes how much entropy is produced inside the system itself. The second law requires that the sum of these two parts be positive, except at equilibrium, when it is zero. But if the system is very far from equilibrium, the first term will be so overwhelmingly positive that *even if the second term is negative*, the sum can still be positive. This means that, without violating the second law, systems far from equilibrium can experience a local entropy decrease. For systems of interest to Prigogine, this decrease manifests itself as a dramatic increase in internal organization. To foreground the connection between self-organizing processes and large entropy production, Prigogine called such reactions "dissipative" systems (Nicolis and Prigogine, 1977). He knew that for Kelvin and other thermodynamicists, this word connoted waste, decline, and death. By using it to denote the spontaneous appearance of organized structure, Prigogine emphasized the important positive role that entropy production can play.

The problem then was to understand how self-organization takes place. Extensive research was undertaken on such chemical processes as the Belousov-Zhabotinskii (BZ) reaction, which exhibits dramatic and highly organized structures far from equilibrium. The primary and secondary reactions involved in the BZ reaction are complex, but the reaction itself is not difficult to produce. All one needs is a petri dish that contains bromate and cerium ions in an acidic medium. The reaction is sufficiently dramatic to warrant describing it as if we were there. What would we see?

Initially the solution appears homogeneous. Suddenly a colored ring forms near the dish's center and begins to pulsate, spreading outward toward the perimeter.[1] We discover that if we jiggle the

[1]The ring is visible because the catalyst for the reaction is a colorimetric indicator. A distinctive feature of the BZ reaction is that the catalyst for the reaction is produced by the reaction itself. When a reaction product feeds back into the reaction in this way, the reaction is said to be autocatalyzing. The feedback loops involved in auto- and cross-catalyzation are characteristic of self-organizing reactions. They provide the mechanisms necessary for the reactions to vacillate between intermediate stages, instead of proceeding immediately from reactants to products.

dish, ruffle our beards over it (as chemists of old used to do), or drop in dust particles, other rings form and fight with the first ring for dominance. Usually the ring oscillating at the highest frequency wins, progressively destroying the waves of lower-frequency oscillators until it is the only one to remain. If we dip a pencil into the dish and "break" a ring, it transforms into a spiral that begins to spin. We find that these fascinating displays do not last very long. As secondary reactions drain the chemicals away from the primary reactions, the displays gradually fade. Setting up the reaction again, this time we stir the reaction. We discover that the continuously stirred reaction creates a display that can sustain itself indefinitely. The stirred solution is originally homogeneous, as the first reaction was. Soon, however, the solution flashes blue, then red, then blue again, at regular intervals. We are seeing a second version of the BZ reaction that produces a so-called chemical clock.

Self-organizing reactions have been known for some time; the BZ reaction was recognized in Russia about thirty years ago by the two chemists who gave their names to it, Belousov and Zhabotinskii. Prigogine did not invent these reactions. Nor did he do the pioneering work of cataloguing their remarkably various forms; the credit for this work goes to Arthur Winfree and others.[2] What Prigogine did was create a metaphysics that places these reactions at the center of a new view of how order emerges. Prigogine demonstrated mathematically that when initial concentrations of the reactants are large, the solution becomes unstable, and local fluctuations appear in the concentration of reagents.[3] Past some critical point, these microscopic fluctuations correlate with other fluctuations spatially removed from them and become the central points from which arise the macroscopic rings and spirals of the BZ reaction. Prigogine and Stengers represent this coordination as macroscopic concentrations "instructing" local regions to engage in a process of self-organiza-

[2]For a review of the BZ reaction, see Arthur T. Winfree, "Rotating Chemical Systems," *Scientific American* 230 (1974): 82–95. Arthur Goldbeter and S. Roy Caplan discuss self-organizing reactions in biology in "Oscillatory Enzymes," *Annual Review of Biophysics and Bioengineering* 5 (1976): 449–73.

[3]An accessible treatment of the mathematics can be found in Jeffrey E. Froehlich, "Catastrophe Theory, Irreversible Thermodynamics, and Biology," *Centennial Review* 28 (1984): 228–251, especially pp. 233–235.

tion through some sort of "communication" between them (1984: 63–89).

Prigogine's explanation is distinct from previous explanations because it does not rely on specific reaction mechanisms, such as stereospecificity, to explain how self-organization takes place. If anything, Prigogine and Stengers overemphasize this aspect of Prigogine's work, implying that it is extraordinary, even miraculous, that millions of individual molecules can "communicate" in such a way as to make macroscopic structures possible. Because they do not distinguish clearly between communication unique to living beings and "communication" between molecules, the rhetoric of *Order out of Chaos* takes on an anthropomorphic quality quite different from what one finds in most articles on chaos.

With this general background, we are now ready to consider what Prigogine and Stengers call the "leitmotif" of their text, the integration of being and becoming. Even a casual reading reveals that the chapter on microscopic irreversibility is crucial to the book's argument. Throughout earlier chapters, important questions are deferred to it; afterward, frequent references are made back to it. The chapter is indeed central, for it undertakes to integrate the timeless world of classical physics with the time-bound world of biology and chemistry. At issue is whether time's one-way direction is intrinsic to reality or an artifact of the observer.

In tackling the question, Prigogine and Stengers attempt to strengthen the case for time's irreversibility by reinterpreting the second law. The second law is involved because it gives time its arrow by defining "forward" as the direction of increasing entropy. Suppose you see a movie in which scattered pool balls magically reassemble into the triangular shape of the rack. You know the film is being run backward, because in real life systems never work like this. Chocolate poured into milk never spontaneously sorts itself into syrup again; ink smudged on a page never flows back again into a single clean line. These results are consistent with the second law's decree that entropy always tends to increase in closed systems. It is important to understand that this tendency is *not* predicted by classical physics. In Newton's equations of motion, negative time can be substituted for positive time and the equations still hold true. As far as Newton was concerned, time could run backward. The same is true for the special theory of relativity; it too treats time as a two-way street. Only the second law puts a point on time's arrow.

For years scientists and philosophers had pondered whether time's forward direction was merely very probable or absolute. The question arose because it was not clear whether the second law was a generalization about what would probably happen or intrinsic to microscopic reality. If the second law was merely probable, then the forward motion of time had an unavoidable subjective element. To see why, consider the pool-ball example again. Once the balls have been scattered from their original triangular formation, for time to run backward they must collide in just the right way so as to form the triangle again. If they form any other pattern (or no pattern at all), from our point of view time would not be running backward. The probability that this one specific pattern will occur is infinitesimally small, because it is only one of millions of possibilities. Although nothing inherent in physical reality causes time to move in one direction, it goes forward and not backward because the probability for events to happen in a myriad of different ways is infinitely greater than for them to happen in one way. And why must things happen in just this way? Because our knowledge defines the triangular pattern, and only the triangular pattern, as "past." Hence the necessity for time to move forward is in this view inherently subjective, an artifact of the observer's presence.

Prigogine and Stengers try to refute this argument by locating irreversibility in microscopic reality rather than in the observer's knowledge. They assert that once the balls are dispersed, the information the triangular form represents has been dissipated. For the pool balls to reassemble, they would have to "communicate" with each other about position, momentum, trajectory, and so on, so that all the different motions would be coordinated in just the right way. Even for a few pool balls, the volume of information involved is very large. If the entire universe were to run backward, it would be essentially infinite. Thus Prigogine and Stengers conclude that time can go only forward because an infinite information barrier divides past from present. In this view the second law is as much a statement about lost information as it is about dissipated energy. Just as some energy is dissipated in every real heat exchange, so some information is lost in every real matter/energy exchange.

To see the force of this argument, imagine that when the balls begin retracing their trajectories, your body experiences a similar reversal of all the processes that occurred during the time you watched the balls disperse. If time could pivot and begin running backward in this way, you would be growing younger while the balls were reassembling. For this to happen, each particle in your body which had been involved in a collision while the balls were dispersing would have to "communicate" with every other particle it hit, so that all of these collisions could now happen in reverse. This would require millions more bits of information than the pool balls needed for their reassembly. If not just you and the balls but everyone else in the world was experiencing a similar reversal, the information required would be unthinkably enormous. Even so, we would have accounted for just a few minutes in the lives of just one species on one smallish planet in one galaxy. For time to run backward throughout the universe, the information would be so staggeringly huge that it is easy to see why Prigogine and Stengers call it an infinite barrier. The problem with time going backward, then, is not just in the observer's knowledge. Rather, it is in the massive correlation of information necessary to have collisions on every level, from subatomic particles to cars on a California freeway to meteorites striking Jupiter, reverse themselves.

To make this argument, Prigogine and Stengers find it convenient to introduce the idea (developed by Ludwig Boltzmann in the late nineteenth century) of correlated and uncorrelated particles. Before two particles collide, they have no necessary correlation with each other. The collision might or might not occur, depending on everything happening around them. The BMW might or might not hit the Honda, depending on whether the Chevette manages to swing around the truck blocking the path ahead of them. But once the BMW has hit the Honda, these two cars *must* be correlated if a time reversal is to occur; and all the collisions that preceded that collision must also be correlated, on every level, back to the beginning of time. When time goes forward there is a role for chance, because small or random fluctuations near a bifurcation point can cause a system to take a different path than it otherwise would. Whether the Chevette misses the truck could depend on whether a butterfly fluttering nearby distracts the driver's attention for a second. But when time runs backward along the same track it took before, every juncture point is already predetermined, and hence chance can play no further part in the system's evolution.

The introduction of chance hints at what is at stake in this argument. Compared to the question whether death is inevitable, the debate over whether the second law is an artifact of the observer or rooted in microscopic irreversibility may seem trivial. Prigogine and Stengers's point, however, is that these concerns are very much related. Humankind has always sought ways to come to terms with its mortality, and one of the ways that figured importantly in the history of science was to envision reality as a timeless realm purged of the irreversible changes that mark human experience. Prigogine and Stengers imply that, on a deep level, this view of reality was motivated by a desire to escape from an existence that seemed all too vulnerable to the vagaries of chance.

Few have felt the appeal of a timeless realm more deeply than Einstein. In his famous mediation on the "Temple of Science," Einstein speculated that the scientist or artist is attracted to his vocation because he is in

> flight from everyday life with its painful harshness and wretched dreariness, and from the fetters of one's own shifting desires. A person with a finer sensibility is driven to escape from personal existence and to the world of objective observing and understanding. This motive can be compared with the longing that irresistibly pulls the town-dweller away from his noisy, cramped quarters and toward the silent, high mountains, where the eye ranges freely through the still, pure air and traces the calm contours that seem to be made for eternity.
>
> With this negative motive there goes a positive one. Man seeks to form for himself, in whatever manner is suitable for him, a simplified and lucid image of the world, and so to overcome the world of experience by striving to replace it to some extent by this image. [Einstein, 1954: 224–227]

It may seem strange that these lines were written by the man who formulated relativity theory. However, Einstein was led to relativity precisely because he saw it as the only way to preserve the overall invariance of physical laws. Before he settled on "relativity," he had considered calling his discovery the "theory of invariance." The appeal that he felt in "the calm contours that seem to be made for eternity" can be sensed when one looks at a Minkowski diagram of spacetime.[4] Because time is one of the coordinates presented for inspection, there is a tendency to feel one is looking at reality from an eternal viewpoint. The illusion speaks to the strong sense Einstein had of a scientific model as a "simplified and lucid image of the world," beautiful in its exactness and clarity.

From this perspective, the assertion that the second law is an artifact of subjective experience is crucial. For if, on the contrary, the second law were intrinsic to reality, then irreversible change would be inextricably woven into the fabric of the universe, and it would be impossible to maintain (or create) an anywhen free from the taint of mortality. The subtext for the debate over the second law is thus a struggle over how one is to see reality on a fundamental level: bombarded by chance, subject to decay, headed toward death; or existing in a timeless realm where chance and irreversibility are merely subjective illusions.

We can now understand the full scope of the intervention Prigogine and Stengers want to make in *Chaos out of Order*. They insist that since irreversible events are demonstrably part of the world, in fact a much larger part than reversible ones, we must recognize that the view of classical physics is at best incomplete. Moreover, they suggest that the price paid for this classical perspective is alienation, for a timeless realm is a world divorced from human experience. In its place they offer a vision that they find both truer to reality and less alienating for the human spirit. The essential change is *to see chaos as that which makes order possible*. Life arises not in spite but because of dissipative processes that are rich in entropy production. Chaos is the womb of life, not its tomb.

There is evidence to support this view. Estimates indicate that the time required for life to arise if it relied purely on random sequencing of amino acids would be far longer than the present age of the universe (Madore and Freedman, 1987: 253). The probability, then, is that some kind of self-organizing process arose from chance fluctuations. The self-organizing process was able to proceed much

[4]Hermann Minkowski, a mathematician, was the first to realize that Einstein's special theory of relativity implied a four-dimensional spacetime. In 1908 he predicted that "space by itself and time by itself are doomed to fade away into mere shadows, and only a kind of union of the two will preserve an independent reality" ("Space and Time," in *The Principles of Relativity* [New York: Dover, 1908]).

more quickly than random sequencing because it possessed the ability to create and replicate form. From this perspective, one could say that the second law in a sense *created* life, for only a world rich in dissipative processes would have been able to support self-organization. To make the second law intrinsic to reality is in this view comforting rather than threatening, for it suggests that the evolution of life was not an accident but a response to the essential structure of the cosmos.

This is why Prigogine and Stengers do not want to concede objectivity to those who, like Einstein, privilege a universe where irreversible change is an illusion. Instead, they imagine a world that is both "objective and participatory" (p. 299). They imply that we can know the world because it is constituted through the same processes that created and continue to govern human life. The passage they quote from Merleau-Ponty about a "truth within situations" aptly summarizes their position:

> As long as I keep before me the ideal of an absolute observer, of knowledge in the absence of any viewpoint, I can only see my situation as being a source of error. But once I have acknowledged that through it I am geared to all actions and all knowledge that are meaningful to me, and that it is gradually filled with everything that may *be* for me, then my contract with the social in the finitude of my situation is revealed to me as the starting point of all truth, including that of science and, since we have some idea of the truth, since we are inside truth and cannot get outside it, all that I can do is define a truth within the situation. [Prigogine and Stengers, 1984: 299]

Merleau-Ponty of course speaks from the perspective of a psychologist and phenomenologist, so it is perhaps not surprising that he would think in terms of a "contract with the social." It is a further stretch for Prigogine and Stengers to associate chemical processes with social reality. Yet their rhetoric is designed to make such a passage easy or plausible. On a deep level, this is no doubt why they tend to write about chemical reactions in anthropomorphic terms. They would make common cause with theorists within the human sciences who challenge totalizing views and celebrate the aleatory and stochastic. Although they approach the subject with a light touch, Prigogine and Stengers clearly agree with Michel Serres that implicit in traditional scientific objectivity is a will toward mastery which feeds on the illusion that the world is separate from us.[5] Only if we blinded ourselves to the complex feedback loops characteristic of self-organizing structures could we think that we could exploit and dominate others without ourselves being profoundly affected by these actions.

These are heady conclusions. And if they are based solely on the scientific data Prigogine and Stengers present, they are shaky ones. Not only does their argument go far beyond the scientific data; it does not sit altogether easily with the data. Recall that the centerpiece of their interpretation of the second law is the infinite information barrier separating past from future. Yet their analysis of self-organizing structures emphasizes the seemingly miraculous way in which millions of molecules can "communicate" with one another. If molecules can engage in "communication" to build self-organizing structures, why can they not "communicate" to cause time to go backward? One could respond, of course, that self-organizing structures have been observed in nature, and that time reversal has not (at least not unequivocally). But this is to place the case on an empirical rather than a theoretical basis, which is precisely the aspect of traditional interpretations of the second law they were trying to overcome.

A further difficulty arises from the way Prigogine and Stengers group together research programs that are quite different in their assumptions, methodologies, and conclusions, as though all of this research points unambiguously toward an order-out-of-chaos paradigm. For example, they mention Feigenbaum's one-dimensional mappings (discussed in chapter 6) without bothering to explain the nontrivial differences that exist between mappings and multidimensional phase diagrams, much less between mappings and the chemical systems that form the basis for their own work (p. 169). Yet connections can be and indeed have been made between self-organizing processes and dynamical systems theory. Roux, Rossi, Bachelart, and Vidal (1980) have written about the BZ reaction in terms of its being a "strange attractor"; Hudson and Mankin (1981) have

[5]Prigogine and Stengers contributed a long commentary on Michel Serres's essay "Lucretius: Science and Religion," in *Hermes: Literature, Science, Philosophy* (Baltimore: Johns Hopkins University Press, 1982).

discussed the chaotic behavior of the well-stirred version of the reaction; and Simoyi, Wolf, and Swinney (1982) have compared the behavior of the BZ reaction to period doubling in one-dimensional maps. However, there is an obvious and important difference between these analyses and Prigogine and Stengers's approach. Prigogine and Stengers emphasize that a large entropy production results in the creation of macroscopic structures where none had existed before. In strange-attractor analyses, the form is considered to be *encoded within the information* the system produces, but these patterns do not necessarily result in stable, self-replicating structures as such.

I want to emphasize that these are criticisms of Prigogine and Stengers's presentation, not of the importance of self-organizing structures. Since *Order out of Chaos* was written (the original French version appeared in 1979), significant advances have been made in understanding how self-organization occurs and how the simpler systems can be modeled. Madore and Freedman (1987) have created two-dimensional computer simulations of the BZ reaction using a cellular automata model that has suggestive implications for some of the larger issues Pirgogine and Stengers raise. It may be helpful to compare and contrast this research with the claims of *Order out of Chaos*, for it illustrates how a more conservative, nonphilosophical approach nevertheless supports the overall thesis about the importance of self-organizing reactions. At the same time, this approach reveals how many links are missing from the causal chain Prigogine and Stengers would forge between the mechanics of self-organizing reactions and a metaphysical privileging of becoming over being.

In the cellular automata approach, the simulation space is divided into cells, and each cell is assigned to one of three states: active, receptive, or quiescent. An active cell can excite an adjacent cell if the adjacent cell is in a receptive state. After a prescribed time, active cells decay into quiescence. After a still longer time, quiescent cells can become receptive again. These sequencing rules mean that active states cannot propagate into areas that are already active, nor can they propagate into quiescent areas.

Suppose that, in the center of the simulation space where all the cells are in a receptive state, a single cell is made active. Excitation will spread outward in a circle from this point. Because active cells decay into quiescence, the cells closest to the point of excitation will quickly become quiescent and will be unable to be activated again immediately. This process provides a "pushing" force that makes the circle into a ring and keeps it radiating outward. If thc ring is disturbed or fractured along its perimeter, spiral arms appear of the form observed in the BZ reaction. Virtually all of the spectacular displays observed in the BZ reaction can be recreated by this simple model (Cohen, Neu, and Rosales, 1978; Duffy, Britton, and Murray, 1980). Recent work has expanded the approach to three-dimensional models of the BZ reaction (Welsh, Gomatom, and Burgess, 1983; Winfree and Strogatz, 1985).

This propagating mechanism may explain how complex spatial organizations can spontaneously come into existence. Spiral forms similar to those observed in the BZ reaction have been found, for example, in nerve axons (Madore and Freedman, 1987), in the lens of the firefly's compound eye (Winfree, 1980), and in slime mold aggregations (Newall and Ross, 1982). This last example is particularly interesting because it is possible to observe the spiral configurations as they form. Slime molds reproduce through spores, each of which is a one-celled organism. As long as there is abundant food in the environment, the mold stays in this form. As soon as the food supply is exhausted, however, the spores swarm together and take on specialized functions, eventually forming a stalk that in turn gives out spores again. A crucial step in the mold's transformation from randomly distributed cells to a complex organized structure is the formation of spiral configurations around the individual spores. Thus it appears that the morphogenic tranformation is achieved through a self-propagating process similar in configuration to the BZ reaction.

A cellular automata model has also been used to explain large-scale spatial organizations, such as the spiral arms of a disk galaxy (Mueller and Arnett, 1976). Of course, many more perturbations are present in galactic systems than in the small petri dishes where the BZ reaction takes place. To account for these perturbations, stochastic elements are introduced into the simulation, giving the spiral arms the "raggedness" characteristic of galaxy formations. The claim of researchers who model systems in this way is not that these formations are all the same, only that they employ mechanisms that are similar in that they achieve a "pushing" force through the interplay among active, receptive, and quiescent areas. The forces responsible for creating these spatial gradients in interstellar gases are obviously very different from those responsible for giving the BZ reaction its characteristic forms. The models thus suggest how morphogenesis can occur, but they do not explain the particular forces at work in a given situation.

As we shall see in chapter 7, this kind of modeling is typical of dynamical systems methods. The procedure raises questions about what it means to model a system, and ultimately about the nature of scientific explanation. The fact that these models have little or no relation to specific mechanisms makes it easier to justify injecting stochastic elements into them, for the absence of a causal explanation makes the intrusion of chance conceivable at any point. Thus Prigogine and Stengers's valorization of chance is implicit in these modeling procedures, although in dynamical systems methods it is presented as a pragmatic technique rather than as a philosophical revolution.

Given Prigogine's interest in the arrow of time, it is perhaps not surprising that he would turn to cosmology, for the fundamental parameters of the universe were no doubt set at its birth. Edgard Gunzig, Jules Geheniau, and he have recently proposed a model that makes the order-out-of-chaos dynamic intrinsic to the formation of the cosmos (1987). This model imagines that at the very beginning of the universe, fluctuations in the quantum vacuum stabilized themselves in the form of small black holes, which in turn became the matter generators necessary to explain why there is something rather than nothing. "As far as energy is concerned," Prigogine and his coauthors write, "the creation of the universe from a Minkowskian [quantum] vacuum is indeed a 'free lunch'" (p. 623). In this model, the scenario Kelvin imagined for the universe is completely reversed. "Heat death" occurs at the beginning rather than the end of time. Thus the transformation of entropy from epilogue to prologue is complete, and chaos is reinstated as our primordial parent.

To see how the model works, it will be useful to explore some of the questions left unanswered by big-bang cosmologies. A big-bang model predicts that residual background radiation would be left over from the explosion. This radiation was found (inadvertently) by Arno Penzias and Robert Wilson in 1965, when they were testing a very sensitive microwave detector for Bell Laboratories and could not get rid of residual noise. No matter in which direction they pointed the detector, the noise remained the same, an indication that it must be coming from outside the atmosphere. Moreover, the radiation remained at the same level, regardless of the time of day or year it was measured. Penzias and Wilson eventually realized that they were detecting the background radiation predicted by big-bang cosmologies. But the extraordinary constancy of the background radiation was a surprise. Repeated observations have confirmed that it does not vary by more than one part in ten thousand. This consistency is striking evidence for the so-called cosmological principle—the premise that the universe is isotropic (that is, does not vary with the axis of observation) and homogeneous.

Why it is surprising that the background radiation should be so uniform? According to the theory of relativity, if light cannot reach a region, nothing else can either, because nothing can travel faster than the speed of light. There would not have been time in the early universe for light from one region to travel to all the other regions, and vice versa. Why then should the background radiation be constant throughout the universe, since the various regions could not have communicated with each other to achieve equilibrium? And why, in contrast to the remarkable homogeneity of the background radiation, should matter be clumped into stars and galaxies?

To answer these questions, Alan Guth and others proposed a "new inflationary cosmology," in which the universe was supposed to have expanded initially at an exponentially greater rate than it does at present.[6] The increased expansion rate of an inflationary cosmology means that a large part of the universe's present dimensions would have been achieved in a mere fraction of a second—that is, in a much, much shorter time than if the expansion rate remained unchanged. Therefore light would have had time to travel throughout

[6]Although the universe is still expanding today, it is doing so at a decreasing rate. One of the unanswered questions of contemporary cosmology is whether the universe will eventually stop expanding and begin to contract. Prigogine's model predicts that the expansion will continue, causing the universe to generate new matter in the far future.

the universe, and thus we have an explanation for why the universe is isotropic and homogeneous.

Although it answered some of the questions left unresolved by the big bang, the new inflationary cosmology raised new questions of its own. Since the expansion rate is presumed to have changed, some mechanism is needed to explain why it slowed down. The explanations that were advanced relied on special circumstances that looked suspiciously like special pleading. These problems led Stephen Hawking, in *A Brief History of Time* (1988), to pronounce "the new inflationary model dead as a scientific theory, although a lot of people do not seem to have heard of its demise and are still writing papers as if it were viable" (p. 132). Among the people who believe reports of its death to be greatly exaggerated are Gunzig, Geheniau, and Prigogine. Ironically, they use ideas Hawking helped develop to overcome objections to an inflationary cosmology.

Hawking points out that although the general theory of relativity leads to predictions of a big bang, the theory cannot be considered a complete explanation of why a big bang would occur, for two reasons. First, it imagines a situation in which particles would be so tightly squeezed together that interactions between them could not be ignored, making quantum mechanics necessary. Second, it concedes that under the kind of compression the big bang imagines, the known laws of physics would break down. Hawking was looking for a way to combine quantum mechanics with relativistic cosmology when he was led to the realization that black holes could emit radiation. In this case, as he puts it, "black holes ain't so black."

Black holes are relevant to big-bang cosmologies because both represent singularities in spacetime. A black hole is formed when a star larger than a certain mass begins to burn out. Without the fuel to maintain itself, the star experiences massive gravitational collapse, and the matter within it becomes extremely dense. This dense matter in turn creates a gravitational field so intense that it creates a singularity in the curvature of spacetime. To visualize this process, imagine pushing the point of a pencil into an inflated balloon (but not so hard that the balloon breaks). The balloon represents the normal curvature of spacetime. The indentation into which the now-invisible pencil point disappears represents the black hole. The pencil point rests on the bottom of the black hole, the point of the singularity where the gravitational field is most intense. On the balloon's surface can be seen an indented circular area whose curvature becomes steeper as it disappears inward. For a black hole, the circumference of this curved area is called the event horizon. Once light or anything else goes past the event horizon, the curvature is so steep that it will never be able to escape. Hence the circumference of the event horizon defines what can happen near the black hole and still remain knowable. Once inside the event horizon, events are intrinsically unknowable because light, and hence any other information, cannot escape the gravitational field to return to an observer.

Hawking introduces the idea that black holes might not be black by asking what would happen if we threw something high in entropy—say, a box full of an ideal gas—down a black hole (pp. 102–105). Would the second law be violated? Clearly, the entropy outside the black hole would decrease. But the second law could be satisfied if we supposed that the entropy inside the black hole increased by a commensurate or greater amount. This supposition implies that it would be convenient to have some way to measure a black hole's entropy—a difficult task, since no direct measurements are possible.

Because Hawking believed that nothing could emerge from a black hole, he had earlier proposed that the event horizon of a black hole could only increase, never decrease. This formulation is reminiscent of the second law's decree that the entropy of a closed system always tends to increase, never decrease. Hawking recounts his feeling of annoyance when Jacob Bernstein, a research student at Princeton, published a paper suggesting that the area of a black hole's event horizon could be taken as a measure of its entropy. In a sense Bernstein's thesis solved Hawking's problem, for it provided a measure of a black hole's entropy. But in another sense it created problems for Hawking's theory. If a black hole has entropy, then it should also have a temperature (since entropy is defined as a function of temperature). And if it has a temperature, then it must emit radiation (otherwise, it could remain at the same temperature forever, in violation of the second law). But if it emits radiation, then it can't be black, can it?

The solution to the dilemma, Hawking explains, was to suppose that "the particles do not come from within the black hole, but from the 'empty' space just outside the black hole's event horizon!" (p. 105). He reasoned that if the space around a black hole were really empty, the gravitational and electromagnetic fields would have to be exactly zero. But in this case their values could be known simultaneously and exactly, a result prohibited by the uncertainty principle.[7] Hence there have to be fluctuations within these fields, which can be thought of as the creation of virtual particle pairs. Unlike real particles, virtual particles cannot exist independent of each other. When they take the form of matter, they come into existence as a particle and antiparticle and immediately annihilate each other. If these particle pairs came into existence at the boundary of a black hole, however, one of them could fall into the hole. Then the other could go free and become a real particle.

In order for a black hole to be an effective particle catcher, it would have to be very small, for the particle pair is separated by only a very short distance. Hawking calculated that a small black hole with a mass of a thousand million tons (equivalent to a large mountain) would have an event horizon diameter of a millionth of a millionth of an inch, about the size of an atom's nucleus (p. 108). Black holes of this diameter are not formed by collapsing stars. They could, however, have been formed at the universe's creation, when pressures and temperatures were sufficiently intense to bring about the required compression. Small black holes differ from large ones in the higher temperature at which they emit radiation. Because radiation energy cannot come into existence from nothing, it must be balanced by a loss of mass within the black hole.[8] As the black hole shrinks, it emits radiation at a still higher energy, and so shrinks even faster. Thus small black holes operate according to a dynamic in which an initially small perturbation drives the system further from equilibrium, which increases the perturbation, which drives it yet further. This accelerating feedback loop means that small black holes could eventually "evaporate." Although no one knows exactly

[7]The strength and location of the field are conjugate variables, and hence are analogous to the position and momentum of a particle, which are more usually mentioned in relation to the uncertainty principle. In its more general form, the uncertainty principle applies to any appropriate pair of conjugate variables.

[8]By Einstein's equation $E = mc^2$, energy is proportional to mass. The radiation streaming out from a black hole is balanced by matter conversion within it.

what happens when a singularity in spacetime pops back out (as the balloon pops back into place when the pencil is removed), the most likely scenario calls for a final burst of radiation as the black hole disappears completely. If primordial black holes do exist,[9] Hawking points out, they would "hardly deserve the epithet *black*: they really are *white hot* and are emitting energy at the rate of about ten thousand megawatts" (p. 108).

With this preamble, we are now ready to understand how Prigogine and his co-authors appropriated Hawking's ideas while changing their drift. They propose a cosmological model in which primordial black holes are created not through the big bang but through the fluctuations in the quantum vacuum which Hawking argued had to be present to keep the uncertainty principle from being violated. Because these black holes have masses even smaller than the ones Hawking imagined (by a factor of ten), Gunzig, Geheniau, and Prigogine call them a "dilute black hole gas" (1987: 623). These extra-small black holes would operate through the same unstable dynamic that Hawking proposed, whereby the black hole generates more radiation and brings more particles into being the smaller it becomes. Hence the clumpiness of matter, which is formed as particles coalesce around the black hole gas. A neat feature of the model is the explanation it provides for why the universe's initial inflationary rate changed. The phase transformation falls out of the equations as a result of the evaporation of the black hole gas! Moreover, the time required for the evaporation matches the time necessary for the universe to expand at an increased rate if it is to be homogeneous. Thus the model explains two features of the present universe which big-bang cosmologies cannot—its isotropy and the uneven distribution of matter within it.

An important difference between this model and Hawking's is that large entropy production is built into the universe from its beginning. As the black holes evaporate, their entropy becomes, in ef-

[9]To date, the existence of such primordial black holes has not been confirmed, despite attempts to locate them. Search strategies have focused on detecting them through the X rays and gamma rays they would emit. Gamma rays do not penetrate the earth's atmosphere; hence a detector must be positioned in space. Moreover, there are other sources of gamma rays around, so it would be necessary to establish that the detected gamma rays came from a single source. These difficulties are perhaps sufficient to explain why none has been observed.

fect, the entropy of the universe. Another distinctive feature of the model is the prediction that as the universe continues to expand, it will eventually reach the point where it is sufficiently attenuated so that the quantum vacuum will again bring more black hole gas into existence, which will in turn generate more matter. Thus, although the universe will continue to expand, it will never reach a point of ultimate dissipation because it possesses the capacity to renew itself. Clearly this model would appeal to Prigogine, for it inscribes the order-out-of-chaos scenario into the creation of the universe. What greater validation of the paradigm than to have it explain why there is something rather than nothing? Not coincidentally, it also weaves the second law into the very fabric of the universe, making it coeval with the universe's formation rather than an artifact added later. Thus Prigogine is able to find in cosmology the answer that eluded him in *Order out of Chaos*.

Such deeply felt considerations as I presume lie behind Prigogine's foray into cosmology do not, obviously, prove the model's validity. But neither do they disprove it. One could argue that significant scholarship in any field is compounded with personal emotion, for why would someone devote to it the countless hours that stretch across months, years, decades, unless it mattered on more than a strictly intellectual level? The fact that Prigogine's work evidently emerges from a vision of the way the universe could or should be, rather than the vision emerging from the work, is reflected in the apparent imbalance between the practical consequences of his theories and the dramatic claims he makes for them. Peter V. Coveney, in a recent review of Prigogine's work which was largely sympathetic (1988), acknowledges the problem.

> A criticism that has not infrequently been aimed at the work of the Brussels School . . . is that, even if it is granted from a philosophical standpoint that irreversibility should be included in the fundamental level of description, there do not appear to be any new and experimentally verifiable consequences to have emerged from the approach. If this were the case, the proclaimed "revolution" in scientific throught arising from the advent of an adequate role for irreversibility and time . . . would hardly be valid.

Coveney, who lists the Université Libre de Bruxelles as one of his affiliations, concludes that he believes the criticism has some merit, but hopes "it will be rectified by consideration of problems of concrete physical interest" (p. 414).

The question of what weight should be given to Prigogine's vision, as distinct from "problems of concrete physical interest," goes to the heart of why his reputation within the scientific community is so uneven. For scientists concerned with the practical problems that make up 90 percent—or 99 percent—of scientific research, Prigogine's work appears to be so much useless metaphysical speculation. From this perspective, his reputation appears to be grossly inflated. But there have been times when vision was central, as in Einstein's quarrel with quantum mechanics, which he pursued in the face of strong evidence against his position because he felt sure that God did not play dice with the universe. Another case in point is Bacon's *Novum Organum*. In spite of being wrong in nearly all of its specific predictions, this text has endured because it had a vision of what science could be. The examples illustrate that at pivotal moments, vision can become the essence of what is at stake, and practical problems can be secondary or even irrelevant.

Disputes about vision do not tend to occur when science is in the "normal" phase that Thomas Kuhn (1970) has identified with the presence of a satisfactory intact paradigm. When such disputes do break out, they often indicate that the prevailing paradigm is under pressure. If the pressure mounts to a crisis, presuppositions can come into view which were invisible during the normal phase. *Order out of Chaos* is an attempt to articulate and refute some of the presuppositions of classical physics. If the mere existence of this book suggests that a paradigm shift is in progress, its mixed reception indicates how far such a shift has yet to go. From talking with scientists who work with complex dynamical systems, I know that many of them regard Prigogine's claims as overstated. Their main interest is in what the new techniques will allow them to do with problems of interest to them. Their involvement with practical problems may partially conceal a paradigm shift from them, for it is often the case that new techniques evolve first, and then attitudes change later as their deeper implications are recognized. I therefore do not rule out the possibility that Prigogine and Stengers's argument may turn out to be correct in its general thrust, even if it is wrong or misleading in its specifics.

The concatenation between a proposed solution to an intellectual problem and a strongly felt personal significance is also evident in Hawking's book. He makes clear that for him, the prospect that the known laws of physics break down at the big-bang singularity is deeply distressing, for he believes that the universe should be knowable by human beings. When he discusses the possibility that the universe is just a "lucky chance," he labels it "the counsel of despair, a negation of all our hopes of understanding the underlying order of the universe" (1988: 133). Later, to introduce the mathematical transformation that he asserts can remove the big-bang singularity, he points out that the "idea that space and time may form a closed surface without boundary also has profound implications for the role of God in the affairs of the universe. . . . So long as the universe had a beginning, we could suppose it had a creator. But if the universe is really completely self-contained, having no boundary or edge, it would have neither beginning nor end: it would simply be. What place, then, for a creator?" (pp. 140–141). Such remarks lead one to believe that Hawking's dissatisfaction with big-bang cosmologies is commingled with questions about God, and that the range of possible solutions is constrained by the kind of God he can accept—namely, a God who agrees in advance to make the universe understandable at every point by human beings.[10]

In this respect Hawking is no different from Prigogine or Einstein. Each is motivated by a vision that underlies and guides his scientific research. How to judge between competing visions? It is often said that the art of scientific inquiry lies in asking questions that can be answered. The general theory of relativity, for example, enabled researchers to ask whether light would be bent by the gravitational field of the sun. When the answer turned out to be yes, it was an important confirmation of the theory's validity.

What this formulation obscures, however, is that the visions that guide theory and research involve issues that are larger than answerable questions can address. Light may be bent by gravitation, but this finding does not answer the question whether irreversibility is intrinsic to reality. Self-organizing processes may take place, but this finding does not answer the question whether chance is to be cele-

[10]I am indebted to Clair James of the University of Iowa for suggesting this idea.

brated. A mathematical transformation that removes the big-bang singularity may exist, but this finding does not answer the question whether the universe is ultimately knowable by human beings. Experimental results can help to shape, refine, and substantiate a vision, but they are not usually sufficient to determine it.

Where does the vision come from? No doubt the circumstances of an individual life play an important role. But visions take hold and spread because they speak to something in the cultural moment. They signify more than the research can demonstrate; and it is this excess signification that produces and is produced by the cultural matrix. Prigogine's vision is of a universe rich in productive disorder, from which self-organizing structures spontaneously arise and stabilize themselves. His vision imagines the cosmos providing, as Alan Guth says, "the ultimate free lunch"—a gourmandizing delight where the pot is never empty, and where the flow of time leads to increasing complexity rather than decay.[11] The vision sees nothingness and somethingness joined in a complex dance, in which vacuums are never truly empty and gaps are never merely ruptures.

To some extent Henry Adams felt the power of this vision, although when he wrote it was only a glimmer of a possibility, and his circumstances led him to cast it in a darker mold than does Prigogine. Stanislaw Lem also shares parts of the vision. More contemporary than Adams and more literary than Prigogine, Lem connects the image of the creative void with poststructuralist theory. If these convergences suggest that the vision is more widely shared than Prigogine's critics acknowledge, they also demonstrate that the values associated with it are more diverse and polysemous than Prigogine recognizes.

[11]Guth's remark is quoted in Hawking, 1988:129.

CHAPTER 5

Chaos as Dialectic: Stanislaw Lem and the Space of Writing

STANISLAW LEM has a formidable range of interests. Trained as a physician, he is seriously interested in mathematics, has taught himself cybernetics, possesses a good working knowledge of biology, knows quite a lot about cosmology, has read extensively in critical theory and philosophy, and writes in Polish, German, and English. The one thing he seems not to know is chaos theory—at least not in the explicit forms discussed earlier, in which chaos is identified with a new set of paradigms. When Lem writes about chaos, he understands it in the older sense of chance, randomness, disorder. Through his reading in cybernetics, he has had ample opportunity to learn about the principle of self-organization that underlies Prigogine's work. However, he has not published anything to date that would indicate he has read Prigogine or knows about applications of the idea of self-organization to chaotic systems.

Yet chaos is a topic of almost obsessive interest for Lem. His writing suggests that he considers accounting for randomness and disorder perhaps the most important issue in twentieth-century thought. Influenced both by his scientific training and by poststructuralism, he is concerned with the interaction between something and nothing as they collaborate to form human meanings. His fascination with disorder, his wide reading in cybernetics, and the influence of postmodern fiction and literary theory have led him to an idiosyncratic blending of some of the same elements that are evident in chaos theory, although the patterns he weaves and the conclusions he draws are uniquely his own. In a sense, he partakes of the foment that led to chaos theory without participating in the paradigm.

Whereas chaos theory identifies a new kind of orderly disorder, Lem sees order and disorder as bound together in a dialectic that enfolds them into each other while still permitting each to retain its distinct identity. Where Prigogine sees the void as a fecund space of creation, Lem envisions it as interpenetrating the apparent fullness of the world, rendering signification problematic and certainty impossible. When Lem's work is seen in the context of chaos theory, it suggests that when certain concepts become highly charged, they will appear in various combinations throughout the culture, not only in the authorized versions that crystallize around new paradigms.

Few writers speak as much about their own and others' works as Lem. His interviews, personal essays, critiques of science fiction, philosophical treatises on language, and sweeping attempts to categorize modern fiction testify to his continuing fascination with the processes and products of writing. In this writing about writing, there is a curious division between process and product—between the way Lem talks about the creation of texts and the way he thinks about fictional works, including his own, once they have been written. Many writers experience a split between the creator and the censor, the intuitive self who experiences writing as subjective creation and the analytical self who stands back from what has been written to revise and reshape the material. But for Lem the division is more radical; creator and censor do not operate within the same space. This does not mean, however, that the dialectic between them is unimportant. It is invisible precisely because it is so important. Displaced from direct articulation, it is the ground from which articulation emerges.

The ratiocinative side of the dialectic is most apparent in Lem's critiques of modern fiction, especially science fiction. Typically he summarizes the plot, identifies the basic ideas, and then shows that they are self-contradictory or incompatible with known facts. The inevitable conclusion is that the work is "twaddle" because it does not satisfy these demands.[1] Although most of these critiques are directed at the work of other authors, his own books are not exempt from scrutiny. He has repeatedly dismissed his early works as flawed or "purely secondary" because the ideas they embody seem to him deficient (1984:93). From such analyses, one might well conclude that for Lem, the cognitive structure of a work *is* the work.

His speculations on the theoretical possibilities open to contemporary writers do nothing to dispel this impression. The strategy he follows in "Metafantasia: The Possibilities of Science Fiction" (1981) is typical. First he gives examples of various kinds of ideas a writer might use as a basis for a story. Then he criticizes the ideas according to their value as rational hypotheses, implying that the bolder and more consistent the hypothesis, the better the work that embodies it. Since it is easy to recognize among these "hypothetical" possibilities ones that he has in fact used in various fictions, it is tempting to conclude that when he writes, he begins with an idea that has been carefully thought out and tested for rational consistency and then proceeds to invent fictional situations that will work out its implications.

According to his own testimony, however, this is far from the case. "Nothing I've ever written," he states, "was planned in an abstract form right from the start, to be embodied later in literary form."[2] Moreover, the works that had relatively more planning he regards as his worst, novels that he acknowledges as his own "only with some discomfort". With the books that he believes are successful, he confesses to have "found [himself], during the writing process, in the position of a reader," as uncertain about what would happen next as if someone else were the author. "Considering myself to be a rationalist, I dislike such confessions," he admits; "I should prefer to be able to say that I knew everything I was doing—or, at least, a good deal of it—beforehand" (1984:94).

The division between his intuitive method of writing and the analytical rigor of his writing about writing has deep roots in his past.

[1]Lem, 1971a:322. See also Lem, 1974a.

[2]For further testimony on Lem's wariness about planning his works in advance, see Istvan Csicsery-Ronay's "Twenty-two Answers and Two Postscripts: An Interview with Stanislaw Lem," *Science-Fiction Studies* 13 (November 1986): 242–260.

In an autobiographical essay titled "Chance and Order" (1984) he speaks of the radical break that occurred in his life when his privileged and secure childhood in the Ukraine was disrupted by the cataclysmic upheavals of World War II. His life broke in two. On one side was a childhood in which he "lacked nothing," indulged by a "respected and rather wealthy" father (p. 88); on the other side was his life during and after the war, in which he "resembled a hunted animal more than a thinking human being" (p. 90). In the first period, life was stable, rational, predictable, but isolated within the closed structure of a family whose affluence made it an enclave within an impoverished and threatened country; after the break, life was unstable, irrational, and unpredictable, but open to the excitement of the streets and the dangers of chance. "I was able to learn from hard experience that the difference between life and death depended upon minuscule, seemingly unimportant things and the smallest of decisions: whether one chose this or that street for going to work . . . whether one found a door open or closed" (p. 90). The image of the open door is significant. His response was not to minimize danger but, in some instances, to court it. He recounts incidents when he deliberately exposed himself to risk, as if to test his ability to survive in a world of chance.

The relation of these experiences to his divided response toward writing was apparently forged during the transition between the two phases. Recalling his earliest writing, in which he would "transport [himself] into fictitious worlds," he emphasizes that he did not "invent or imagine them in a direct way. Rather, I fabricated masses of important documents when I was in high school in Lvov: certificates; passports; diplomas . . . permits and coded proofs and cryptograms testifying to the highest rank—all in some other place, in a country not to be found on any map." Wondering if he invented these documents to assuage an "unconscious feeling of danger," he concludes that "I know nothing of any such cause." Yet the fact remains that at this critical juncture, when the order of his early life was threatened by events beyond his control, he turned to the invention of texts as a way to create a space that would be open to uncertainty and danger, but within which he would be protected by the "full power of authority." It is no exaggeration to say, as he does, that the two opposed principles of chance and order "guide my pen"

(p. 88). Indeed, one might conclude that it is this dialectic that put the pen into his hand.[3]

For Lem, writing creates a mediating space in which openness and closure, chaos and order, creation and ratiocination engage each other in a self-renewing and multilayered dialectic. When writing is in progress, the emphasis is on the open-endedness of the process—anything may happen, infinite possibilities lie open to the creator, chance fluctuations create situations whose outcomes cannot be foreseen. But this openness also undermines authority, putting the writer "in the position of a reader," subject to vicissitudes he cannot control (1984:94). He can reassert order by writing about writing, creating documents whose purpose is not primarily creation but regulation (recall the passports, diplomas, certificates that were among Lem's earliest creations). In this writing about writing, possibilities are enumerated only to be evaluated by the rational faculty and, as often as not, dismissed. Because writing about writing unfolds in a separate space from writing as such, the dialectic that engages the creative and ratiocinative self is displaced from direct articulation. On this level, there appears to be simply a dichotomy, rather than a dialectic, between the two modes.

The dialectic begins to come into sight when we view both the writing and the writing about writing as, after all, writing. When writing is ongoing, it creates an open space rich with latent, unarticulated possibilities. Yet writing, especially successful writing, produces products—pages, drafts, texts, books. Once these products exist, the space has been closed, the order of the words stabilized. How to maintain the tension between chance and order which is for Lem the condition of writing, even though the writing itself has ceased? In his writing about writing, he suggests a host of possibilities—for example, interspersing random "noise" throughout the text so that one can never be sure what is message, what chance. "If we posit that the task of literature is not to ever give a definitive explanation of what it presents, and is therefore to affirm the autonomy of cer-

[3]The conjunction of World War II and Lem's fascination with chance suggests that the upheavals of World War II may have been important factors in focusing the attention of that generation on chance. The war effort was intimately bound up with the development of information theory, which was created in part to control chance situations.

tain enigmas rather than to enter into explanations, then the most engimatic of possible secrets is a purely random series" (1981:65). Here the value of chaos for Lem becomes clear; it allows literature to elude criticism's attempt to close the space of writing. Yet in another sense, in this passage he is in fact attempting the "definitive explanation" that he hopes chaos can defeat. Driven or led to ratiocination, he yearns to create a chaos that can defeat explanation; faced with chaos, he cannot resist entering into explanations. The effect of these convoluted crossings within the space of writing is to create a dialectic that ceaselessly renews itself, wresting rational explanations from the enigmatic silence of the text even as it opens fissures within the text which subvert those same explanations.[4]

In Lem's view, the dialectic between chaos and order occupies a central place in all writing. Far from seeing it as an idiosyncrasy peculiar to him, he believes it is intrinsic to language. On the one hand, language is "the instrument of description," operating to bring order and stability to perception; on the other hand, it is "also the creator of what it describes," caught in a self-referential loop whose lack of ground leads to radical indeterminacy (1981:62). Lem is of course not alone in emphasizing the self-referentiality of language. Julia Kristeva sums up the spirit of a generation when she writes about the "theoretical ebullience" of criticism following its "discovery of the determinative role of *language* in all human sciences" (1980:vii). Yet there are important differences between poststructuralists such as Kristeva and Lem's version of that "theoretical ebullience." While he would agree with Kristeva about the importance of language in the construction of culture, he is also interested in informational markers that distinguish between description and creation. He believes, in other words, that language can be referential as well as self-referential. Because creation through language is for him a double act, transforming stochastic process into ordered text and textual order into chance events, his writing has affinities with poststructuralism but is perhaps distinct from it in its insistence on the continuing tension inherent in the dialectic. Acknowledging the void, he also posits a "somethingness" that remains apart from the postmodern vacuum.

[4]Frank Occhiogrosso, 1980, discusses Lem's *The Investigation* as an example of a permanent enigma.

In addition to his view that language "always has meanings oriented toward the world of real objects"(1973b:26), Lem is distinguished from other poststructuralists by his belief that literature must be about something other than textuality if it is to engage ethical questions. The disappearance of "conventional structures of ethical judgment all over the planet" is in his view profoundly bound up with the kind of space in which the modern writer works (1981:61). In "Metafantasia" he argues that the energy for literary creation has historically come from the writer's resistance to conventional ethics. If there are no conventional ethics, the writer faces a very different kind of problem: he or she must then create out of a void. The erosion of traditional values has turned the writer's task on its head. The contemporary writer must *invent*, as well as overcome, the resistances on which writing depends. He must (like Lem) conceive of writing as a double act, at once introducing the constraints that will carve a defined space out of the void and opening fissures within it which will protect the space from oppressive closure.

The metaphoric interplay between enclosed and open spaces is thus central to Lem's thought, whether the subject is culture, ethics, or language.[5] The project of his writing is to define a space that is neither full nor void, excessively cluttered nor empty, because these extremes threaten to destroy writing. Nevertheless, it is only by articulating the extremes that one can create the space of writing, for they are the boundaries that delineate it. In the face of this double jeopardy, Lem's characteristic response is to turn the extremes back on themselves. By enfolding openness into closure, chance into necessity, excessive information into empty language, his writing creates a space that seems to exist apart from the boundaries that bring it into existence. Without the dialectic the space of writing as he envisions it could not exist, for in a very real sense the dialectic *is* that space. The words on the page are the visible trace that the dialectic leaves behind. But for him the dialectic "guid[ing his] pen" is writing-in-itself.

The operation of the dialectic can be illustrated through two representative works, *The Cyberiad* (1974b) and *His Master's Voice*

[5]Jerzy Jarzebski also comments on the marked division between Lem's rationalism and his "special inclination for phantasmagoric visions," a dichotomy that Jarzebski suggests can be modeled through "open and closed space" (1977:123). I am indebted throughout this chapter to Jarzebski's insights.

(1983). *The Cyberiad* is a collection of cybernetic fairy tales written in a grotesque mode. *His Master's Voice* is a psychologically realistic novel that is arguably the best of Lem's serious fiction. Operating within different generic conventions, the two texts are not so much opposites as mirror reflections of each other. The same dialectic is at work in both, although it proceeds in opposite directions.

Consider first *The Cyberiad.* The stories in this collection are noteworthy for their absence of a reality principle. Any idea, no matter how absurd, can be used if it can be put into words. How does one materialize a dragon? By amplifying its probability. How does one answer ultimate questions about the meaning of life? By building a Gnostotron that knows everything. Creation takes place because the appropriate words exist or can be invented, not because objects to which the words refer exist. In view of this linguistic zaniness, it is hardly surprising that the central problem in *The Cyberiad* is how to create something from nothing. When everything is possible, the urgency of generating internal constraints is especially pressing.

His Master's Voice, by contrast, unfolds as the autobiography (or "antibiography") of the preeminent mathematician Peter Hogarth. As its foregrounding of science suggests, *His Master's Voice* has a much firmer internal reality principle than *The Cyberiad.* Along with this difference in initial assumptions goes a corresponding difference in goals. Characters in *His Master's Voice* are concerned not with creating from the void but with determining the status of something that has emerged from the void. This is one of the cases where Lem's "hypothetical" suggestions "retrodict" books he has already written, for the neutrino transmission (the voice of the title) is so densely coded with information that it is virtually indistinguishable from noise. For the characters as well as the reader, the transmission represents a permanent enigma (about which Lem had speculated in "Metafantasia"). As a result, the claustrophobic enclosures of the text are opened to radical uncertainty.[6]

It is now possible to see how the two texts reflect and invert each

[6]Jarzebski (1977:122) comments on how marked Lem's work as a whole is with a sense of claustrophobia, "by the suspicion that the apparently limitless universe surrounding us is in reality a huge, misleading stage backdrop designed to fit our cogni-

other's assumptions in the mirror relation of their dialectics. The emptiness of *The Cyberiad* is introduced as a premise, an absence of originary essence, whereas the emptiness of *His Master's Voice* is presented as a conclusion, a result of informational excess. *The Cyberiad* foregrounds the emergence of its language from the void, emphasizing its creation *ex nihilo*. It then introduces successive constraints until ethical judgment finally becomes possible. *His Master's Voice*, by contrast, appears initially to validate its language as a referential symbol system, locating its subject in reality rather than in language. But this referentiality is decentered by the neutrino transmission, whose excessive signification gives rise to multiple (and multiplying) interpretations. At first ethical judgment seems clear-cut, even complacent. But by the end, judgment has been so contaminated by hermeneutics that it is only by an act of faith that the narrator can achieve any kind of closure at all. Beginning at opposite ends of the open/closed spectrum, the dialectics meet in the middle. Whether the text begins with order or chance, referentiality or self-referentiality, by the end both polarities have been so enfolded into each other through the operation of the dialectic that they are inextricable.

Reinforcing this mirror relation are different stances toward the language that creates and the language that describes. *The Cyberiad* introduces semantic markers that distinguish between descriptive and creative language by introducing a complex set of narrative/frame distinctions. As its narrative structure grows increasingly hierarchical, ethical judgment emerges through implicit connections between different narrative levels.[7] In *His Master's Voice*, the basis for ethical judgment is articulated at the beginning but becomes progressively more ambiguous as the narrator's descriptions of his life, scientific work, and colleagues engage each other in circular reasoning that makes it impossible to contain cause and effect within a logical hierarchy. Whereas in the grotesque work ethical judgment

tive faculties, while the essence of things is hidden away somewhere else, behind the stage." When queried about the Jarzebski comment in the Csicsery-Ronay interview, (see n. 3), Lem replied, "As for claustrophobia, I don't feel any . . . but I can conjure an aura of claustrophobia or agoraphobia, if needed, in my writing." Whatever Lem's personal psychology, claustrophobia is a prominent feeling in his writing.

[7]For a different view, see John Rothfork, 1977.

emerges through the introduction of formal structure, in the serious novel it grows more problematic as the initial regulating structures prove insufficient to order the text. By the end of both texts, the paradoxical configuration of the textual space renders ethical judgment both inescapable and irreducibly complex.

The dialectic is already in operation at the beginning of *The Cyberiad*, when Trurl creates a machine that can make anything that starts with *n*. To test it, Klapaucius asks it to make Nothing, whereupon the machine starts unmaking the world. This opening story demonstrates how Lem's dialectic simultaneously invents constraints and deconstructs them. In construing "Nothing" as a substantive noun that has positive being-in-the-world rather than negation describing a lack of being, the creation machine performs a linguistic sleight-of-hand that transforms the void into the full. At the same time, to create this fullness the machine must deconstruct everything in the world, thus turning it into a void. By enfolding the void and the full into each other, this opening story creates through writing the space of writing.

Central to this paradoxical enfolding is the balance between creation and destruction. To prove it can make anything that starts with *n*, the machine must continue to make Nothing until it finally deconstructs everything, including the constructors and itself. Fortunately, the constructors are able to deflect it from this course of uncreation before deconstruction is complete. Some things still exist, although the world remains "honeycombed with nothingness" (1974b:8). The machine has operated long enough, however, to undermine the referentiality of language. For those objects it has deconstructed, only names remain to signify the resulting voids. "The gruncheons, the targalisks, the shupops, the calinatifacts, the thists" have all disappeared as objects in the world and remain only as signs within the text (p. 6). The ontological security of realism has thus been undercut, since language is revealed as an instrument of uncreation as well as of creation. However, to deconstruct everything would be self-defeating, not least because for Lem creation depends on a dialectic between something and nothing. If there were only the void, creation would be impossible. In this just-so story explaining how the world came to be as it is, the partly deconstructed universe resembles a Swiss cheese, for within the heavens dotted with stars gape holes left by the annihilated "worches and zits." The universe has become the space of writing, containing both self-reflexivity and representation in its paradoxical configuration.

Within the space opened by this initial writing the rest of the text unfolds, evolving through a continuing textual interplay between emptiness and excessiveness. In the story of the Multitudians, for example, Lem imagines a culture in which numerosity is everything. Only masses are valued, not single beings. The uses to which the King of the Multitudians puts his subjects are significant. In an attempt to prove to Trurl the delights of numerosity, the King points out that his subjects can literally be made into signs—living mosaics "providing sentiments for every occasion" (1974b:180–181). Trurl, unimpressed, thinks to himself that "an overabundance of thinking beings is a dangerous thing, if it reduces them to the status of sand" (p. 180). The King's suggestion that his subjects can be used rather like Hallmark greeting cards links excessive numbers to a textuality that empties the subject of its traditional humanistic content. In Lem's view, if language is empty, it cannot be used to express ethical concerns. If the subject is empty, it cannot be used as a locus of value. The planet of the Multitudians thus presents to the constructor the same erosion of traditional values that Lem sees as characteristic of the modern era.[8]

For Trurl, the problem is more practical than philosophical. Although he dislikes the King's ethics, he dislikes still more his suggestion that Trurl accept "two or three hundred thousand" of his subjects as recompense for creating a Perfect Adviser (p. 180). Trurl insists on gold, and the King reluctantly agrees. When the King later attempts to defraud Trurl of his rightful payment, Trurl devises a stratagem to make the King fall victim to the very superfluity of signs that he tried to urge on Trurl. The stratagem is simplicity itself. Trurl sends an innocuous letter to the Adviser, knowing that the suspicious King will believe the letter is written in code. Because there is no code, there are no constraints limiting interpretation, and the King's experts generate more and more possible messages from

[8]Rothfork (1981), discussing the ethical implications of "having everything," places Lem's view of superfluity in the context of utilitarianism.

the letter. The King, having ignored ethical constraints in turning his subjects into signs, is now confronted by a textual void opened by the lack of constraints within Trurl's letter.

Determined to arrive at a unitary meaning, the King calls in the world's "greatest expert in secret writing," Professor Crusticus. The professor, the "distinguished discoverer of invisible sign language," assures the King that he can find the text's one true meaning. He discovers that if he "added up all the letters of the letter, subtracted the parallax of the sun plus the annual production of umbrellas, and then took the cube root of the remainder," he would have the answer. The King's agents confirm the professor's solution by torturing the unfortunate citizen whose name is an anagram of it (p. 191). What the professor's solution exposes is not, of course, the putative answer that the King seeks, but a moral of which the King remains oblivious, although it is clear enough to the reader. Theories about language which claim that it is free to be interpreted in any way whatsoever are the allies and precursors of state terrorism.

The connection between the textual politics of empty language and political violence is underscored when Trurl too is ensnared within the proliferating signs. By encoding the Adviser's program with his own wisdom, Trurl is put in the difficult position of "attempting to conquer himself, for the Adviser was, in a sense, a part of him" (pp. 187–188). Trurl is vulnerable to even such an oaf as the King because he has reduced his identity to a series of signs and replicated them within the Adviser. If identity is merely a collection of signs, capable of dissemination through multiple cybernetic texts (a possibility explored at length in "The Mischief of King Balerion," another story in *The Cyberiad*), then on what basis can human rights, which rest on the sanctity and uniqueness of the individual, be justified? The question is at the center of this text, as it attempts to construct spaces in which the chaotic superfluities of disseminating acts and words are balanced against the oppressively closed structures of tyranny. When the balance shifts too far in either direction, the results are invariably disastrous.

The potential for violence that opening identity to dissemination can unleash is brought forcibly home for Trurl in "A Good Shellacking." In this story, Trurl attempts to trick his fellow constructor, Klapaucius, by passing himself off as "a Trurl," a copy of himself which his marvelous "Machine to Grant Every Wish" has created. Taking the "copy" at its word, Klapaucius locks it in the basement and starts to beat it. Trurl howls in protest, but Klapaucius insists that he has every right to use Trurl as he pleases, since he is only a copy, not the real Trurl. Finally Trurl escapes and, when Klapaucius calls the next day, tells him that he was forced to dismantle the false Trurl because it defamed his good friend Klapaucius. The shrewd Klapaucius knows perfectly well, of course, that there never was a copy. What Trurl has dismantled is a sign, not a physical object.

When the copy is a cybernetic organism rather than a sign, the problems caused by proliferating identities are not so easily solved. Significantly, when Lem's characters are confronted with the perplexing issue of distinguishing between an original and a created self, they often find themselves unable to impose *any* ethical solution upon the situation. In "How Trurl's Perfection Led to No Good," for example, Trurl creates a miniature kingdom in a box and peoples it with tiny beings as a toy for the tyrant Excelsius. When Trurl returns home, Klapaucius points out to him with horror that because the creatures Trurl has made are in every way indistinguishable from real beings, they are real: a "sufferer is one who behaves like a sufferer" (p. 169). Because of Trurl, living creatures who are trapped within the box are entirely at the mercy of Excelsius. Meaning only to placate Excelsius, Trurl has condemned them to a living hell. Creating the box was clearly a mistake; how to correct the error? Destroy the box? Professor Dobbs, encountering similar difficulties in "Non Servium" (in *A Perfect Vacuum*, 1971b), finally decides that he has no ethical options. The best he can do is to delay for as long as possible throwing the switch that will kill his artificial beings.

Trurl's dilemma has a happier outcome, for when he returns to the kingdom-in-a-box, he finds that the miniature beings have escaped their enclosure and turned the tables on their tyrannical ruler, exiling Excelsius to the void. Clearly a startling reversal has taken place, for one of the text's most claustrophobic and tyrannically oppressive enclosures has been opened, while the tyrant has been transformed into a rather picturesque moon. How did this marvel occur? Although no explicit explanation is given, there are hints that it comes about because the inhabitants of the box are capable of self-organizing processes. If ethical solutions cannot be imposed from without, perhaps they can arise from within.

This conclusion paves the way for the reemergence of ethics from the space of writing. While the stories have been showing various kinds of self-organizing processes in action within the narratives, the narrative space itself has grown increasingly complex and hence susceptible to self-organization in its own right. In Lem's view, literature—indeed, language itself—is engaged in a feedback loop in which articulating an idea changes the context, and changing the context affects the way the idea is understood, which in its turn leads to another idea, so that text and context evolve together in a constantly modulating interaction. Regulating structures provide a way to control this interaction and use it constructively.

In general, the hierarchical structures characteristic of self-regulating systems evolve because they are necessary, and they are necessary because they introduce the constraints that bring feedback loops under control by partially "insulating" the changes going on at one level from those at another. How these self-organizing processes occur is illustrated through the increasing complexity of *The Cyberiad*'s narrative structure. From the first story on, feedback loops are created which enfold the narrative space into itself. Consequently, as the stories progress, the space becomes more and more convoluted. By the "King Genius" story, no fewer than three distinct narrative frames are employed, with as many as fifteen stories nested inside a single frame. The structure implies that the narrative/frame hierarchies must be this complex to control the proliferating narratives. At the same time, the greater complexity leads to the possibility of multiple interactions between levels, so that the space is opened to narrative proliferation again. As chaos leads to order, and order back to chaos, the narrative comes to resemble an organism that grows by periodically dissolving and reassembling, each time at a higher level of complexity. In this sense the narrative is a cybernetic organism, manifesting within itself the same self-organizing processes that the stories take as their subject.[9]

It is this self-organization that allows ethical judgment to be in-

[9]For a subtle analysis of how Lem's novels in general become "cybernetic," see Robert M. Philmus, 1986.

trinsic to the space of writing, emerging from within rather than imposed from without. As the text undergoes periodic deconstruction and reassembly, the feedback loops that originated the self-organizing processes work to increase ethical awareness. In one of the "King Genius" stories, for example, we are told about a book by Chlorian Theoreticus the Proph titled *The Evolution of Reason as a Two-Cycle Phenomenon*. According to the Proph, robots and humans are joined by a "reciprocal bond" of mutual creation (p. 244). First humans evolve and create robots. In time the humans die out, and the Automatas, "having freed themselves from the Albuminids, eventually conduct experiments to see if consciousness can subsist" in protein (p. 245). So the robots create "synthetic palefaces," who in turn eventually again make robots. Because the two phases are bound together by a continually evolving feedback loop, neither human nor robot can logically claim to be "original." Conversely, neither one is merely a copy to be used for another's pleasure. The scenario implies that in a cybernetic age, although neither silicon mechanism nor protoplasmic organism has priority, both have rights. The feedback loops that configure the space of writing thus also serve as a model for ethical awareness.

So complex are the multiple interactions between stories and between different narrative levels within the same story that it would take hundreds of pages to explicate fully the connections between ethics and language which emerge in this little book. As an example of this complexity, consider the "Doctrine of Inaction," first proposed by King Genius, the auditor of Trurl's stories. Reasoning that if one cannot control all of the consequences of one's actions, it is better not to act at all, King Genius has abdicated his throne and fallen into a deep depression from which only the most entertaining stories can rouse him. In this context, the Doctrine of Inaction appears as moral paralysis. "Inaction is certain, and that is all it has to recommend it," Trurl admonishes the King. Unlike the King, Trurl favors action precisely because it is uncertain. "Therein lies its fascination," he asserts (p. 177). Trurl's judgment is not the last word on the subject, however, as the subsequent story "Altruizine" makes clear. In a different context, the uncertainty of proliferating actions can be more devastating than the certainty of doing nothing.

In "Altruizine," Klapaucius moves out of the King Genius frame to continue his quest for the perfect action inside another story told by Bonhominus to Trurl. Bonhominus recounts how Klapaucius is given altruizine by HPLDs, the beings who represent the Highest Possible Level of Development. The HPLDs have been convinced, through long and unhappy experience, that the only sound policy is the Doctrine of Inaction. Consequently, they present altruizine to Klapaucius by way of an object lesson. The drug has the peculiar property of causing everyone to feel what everyone else in the immediate vicinity is experiencing. Klapaucius presents it in turn to Bonhominus, who is convinced that he can use it to solve the world's ethical problems. He subsequently dumps it in a town's water supply. Without the controlling constraints imposed by hierarchical structures, the feedback loops that the drug initiates quickly spiral out of control. A man has a toothache, for example, and his neighbor runs screaming into the street. The chaos that Bonhominus unleashes is linked through the feedback image with the relatively simple narrative/frame structure that this story employs. The implication is that ethical judgments appropriate to the convoluted space of "King Genius" may not be appropriate to a space that is less complex and hierarchical. The way the space of writing is configured is thus intimately bound up with the kind of ethical judgments that are appropriate within it.

The "something" that emerges from the void in *The Cyberiad* is not, then, a general rule of conduct, but a dialectic that circles back on itself to raise the level of ethical awareness, slowly and often painfully. That there finally are no definitive answers is a foregone conclusion, given the complexity and diversity of the ethical questions that are posed. There are, however, correlations and symmetries that reveal how language and ethics are interrelated. Ethical awareness for Lem means being aware of these relations. By enfolding the full into the void, openness into closure, the dialectic has caused ethical awareness to reemerge from the vacuum of a deconstructed world.

His Master's Voice is like *The Cyberiad* in its refusal to validate ethical standards that hold true for any time, any place. The void that this serious novel confronts, however, is essentially different from the linguistic emptiness that underlies the grotesquerie of *The*

Cyberiad. More committed to representational language, *His Master's Voice* is also more serious about creating psychologically plausible characterizations. The void here has two loci: the deep vacuum of space, from which the neutrino transmission emerges; and the deep recesses of Peter Hogarth's personality, from which the language that constitutes his narrative emerges. In Hogarth's account as well as in the letter from the stars, the dialectic operates to produce out of a vacuum information so excessive that it finally cannot be distinguished from emptiness. The problem, then, is not how to recuperate ethical judgment from the void, but how to protect it from proliferating significations that threaten to bury it under their excessiveness.

The problem is introduced in Hogarth's "Preface," where he breaks scientific decorum to tell his readers about the dark underside of his public image. Calling his revelations an "antibiography" because they conflict with the beneficent image his official biographers have created (p. 18), Hogarth asserts that his dominant personality traits are malice, cowardice, and pride (p. 16). The catalyzing event that revealed him to himself was his mother's death, which he witnessed as a child. Seeing her die slowly, agonizingly, he concluded that evil was stronger than good. In that case, his cowardice prompted him to side with maliciousness rather than benevolence. As an adult, Hogarth attempts to construct a personality directly opposite to these innate tendencies and so pardoxically reveals to himself his last innate trait, pride (pp. 6–8).

But then it occurs to him that death may be the work not of evil but of chance: millions of chance accidents, chance misreadings in the cells that accumulate as the years go on. He devotes his talent to proving that chance has played a determining role in human nature. "With the formulae of stochastics I strove to undo the evil spell" (p. 10). Now comes a subtle reversal. So successful is this brilliant mathematician at his chosen task that chance is indeed established as the determining factor in human evolution. Paradoxically, then, chance no longer opposes necessity but *becomes* necessity. Thus a new rebellion is necessary: "I know that Chance fashioned us, put us together as we are—and what, am I to follow submissively all the directives drawn blindly in that endless lottery?" (p. 18). If his innate personality is the result of chance, Hogarth will seek freedom by fighting against this "endless lottery." Thus he strives to become what he naturally is not—benevolent, brave, humble.

In Hogarth's antibiography, a familiar pattern is at work.[10] The dialectic circles back on itself so that two opposites, chance and necessity, are enfolded into one another. Whereas in *The Cyberiad* this circling back caused ethics to emerge, in *His Master's Voice* it undermines the initial clarity of Hogarth's ethical judgments. For example, what if someone were naturally good? Would it follow that to attain freedom, he should strive to be evil? Though Hogarth poses the question, he is unable to give a satisfactory answer. Once articulated, the dialectic seems to take on a life of its own and will continue to dominate the events not only of Hogarth's life but also of humankind's encounter with the letter from the stars. Ethical judgments that begin by being clear-cut become progressively ambiguous as the dialectic proceeds.[11]

Already at work in Hogarth's "Preface," the dialectic continues to enfold chance into necessity through the dramatic story of how the "letter" is discovered. Swanson, a physicist who is part guru, part scientist, hits upon the idea of using recordings of cosmic background radiation to generate random number tables. Random numbers are useful in scientific work because the sequences within them manifest no consistent pattern; they may thus be used to simulate chance events. A scientist using the random number tables discovers, however, that sequences in the first volume are repeated in the second. Concluding that Swanson copied his results from the first volume into the second rather than transcribing new radiation recordings, he sues Swanson for fraud. But Swanson has not cheated. Across a narrow band, the neutrino radiation does in fact repeat itself about once every two days. This repetition is all that enables

[10]Hogarth's circuitous reasoning in the "Preface" is strikingly parallel to that of Jacques Monod in his *Chance and Necessity* (1972). Monod's account, like Hogarth's, emphasizes the interplay between chance permutations and genetic replication in the evolutionary process. Published in France a year after *His Master's Voice* appeared in Poland, *Chance and Necessity* is unlikely to have been a direct influence on Lem, although Monod's technical articles may well have been part of Lem's reading. In any case, Monod's text provides a context that clarifies Lem's novel.

[11]Edward Balcerzan, 1975, makes this point in a different context.

the radiation to be recognized as a message, for it is so densely coded that in all other respects it is virtually indistinguishable from noise. The government turns the investigation of the transmission into a top-secret project—"His Master's Voice."

Despite the best efforts of the project's scientists, they never discover whether the radiation is in fact a letter from the stars, or indeed if it is a message at all. Hogarth sees it as a "particularly complex Rorschach test" in which the scientists "attempt, behind the veil of incomprehensible signs, to discern the presence of what lay, first and foremost, within ourselves" (2:32). First seen as noise, then as signal, the transmission also acts as a mirror, reflecting the systematic misreadings of what Hogarth calls Earth's "predatory" science (13:151).

Almost alone among the scientists, Hogarth realizes that attempts to interpret the letter can never escape the hermeneutic circle. The reflexive nature of the project becomes apparent when an early decision is made to use binary code. Later Hogarth notices tiny discrepancies in the transmission, so minute that they cannot be translated into binary code. When his fellow scientists argue that the discrepancies represent a slight blurring of the transmission as it travels through space, Hogarth proves that the transmission is accurate to the limits of the recording instruments. The transmission's code, in other words, is not binary. But the scientists continue to act as if it were, thus encoding their own assumptions into the presumed decoding of the letter. Although Hogarth does not dwell on the matter, the incident illustrates how chance enters into the translation and becomes fossilized there as initial hypotheses become unexamined premises. What the scientists see in the transmission, Hogarth more than hints, is what they have projected onto it.

The scientists are doomed to misunderstand because they can never escape their anthropomorphic perspective, encoded not only into the symbol systems but into the very structure of their brains. In that case, the possibility arises that the mistranslations are in some way *systematic*, not the work of chance but the necessary consequence of human brain structure and evolution. Since this evolution itself originated in chance, however, there exists no possible filter by which the systematic distortions of necessity can be separated from the fortuitous operations of chance, even if one assumes that such a transcendent, nonhuman viewpoint is possible. So Hogarth, and all of the other scientists, are trapped in what he calls "carousellike" reasoning (2:28) in which premises become conclusions and conclusions premises, chance necessity and necessity chance. Beginning with chance, the dialectic moves toward necessity, then circles back to enfold both within the same space.

The dialectic is also at work within Hogarth's interpretation of his antibiography. Does he reveal the secrets of his private life to illuminate his actions in the project? Or does he undertake to write the project's history to indulge in narcissistic self-revelation? Recognizing the possible circularity, he wants to believe that the "ugliness of my malice I made public . . . to divorce myself from it." But suppose that "by stealth, it penetrated, permeated my 'good intentions' and all the time guided my pen. . . . In this diametrically opposed view of the matter, what I held to be an unpleasant necessity . . . becomes the primary motive, while the subject itself—His Master's Voice—is a pretext that comes conveniently to hand" (2:28). In the same way that chance and necessity are enfolded into the same space within the antibiography, so these two texts are further enfolded into each other as pretext becomes text, text pretext. The dialectic that Lem saw "guid[ing his] pen" applies as well to Hogarth and the narrative spaces his writing creates.

The dialectical circularity of the novel is reinforced by its posthumous status. Because Hogarth dies before his book is published, it is the editor, Thomas Warren, who decides that the antibiography will serve as preface rather than as epilogue to Hogarth's account. The arrangement implies, of course, that the scientific work grows out of Hogarth's interpretation of his life. A note from Warren explaining why he chose to put the antibiography first makes us aware, however, that the book could just as easily have been assembled the other way around. Is the order of the text finally determined by necessity or by chance? We have no way to know, since both text and pre-text have already been caught up in a circular dialectic that makes it impossible to establish which came first.

The dialectic circles endlessly because at the center of the text is a permanent engima: the neutrino radiation. As long as the transmission remains incomprehensible, neither Hogarth nor anyone else can close the narrative space by showing which of the data foreshadowed the known end. The transmission thus acts like a hole puncturing the textual enclosure of the novel. Painfully aware of the openness that this enigma introduces into his text, Hogarth attempts to compensate by including as much as possible within his narrative. "I do not know what it is among the people of the Project that determined finally the Project's fate," he admits. "Therefore, just in case, thinking of the future, I am also presenting here those bits and pieces that I have not been able to put into any coherent whole. Perhaps someone else, someday, will manage that" (4:57). The hope is illusory, for as Hogarth himself acknowledges in his attempts to decode the letter, it is impossible ever to describe anything completely. Any description presupposes a frame of reference that limits, even as it creates, what is said.

The hope that a definitive reading may one day be possible, illusory though it is, impels Hogarth to write his antibiography and to include in his account personal, often unflattering portraits of his colleagues. How to locate in this complex space the relation between cause and effect which readers ever since Artistotle have demanded of fictions?[12] The difficulty is not merely that Aristotelian assumptions of unity do not apply. They also do not necessarily *not* apply, for it may be that most, or even all, of the data are relevant. But we have no way to know. Because the transmission remains an enigma, it is impossible to read the text teleologically and consequently impossible to distinguish chance events from necessary outcomes.[13] There is no possibility for a definitive interpretation of the text, whether the text is conceived as the neutrino transmission, Hogarth's antibiography, Hogarth's account of the project, Thomas Warren's edited text of Hogarth's accounts, or the entirety that is Stanislaw Lem's novel.

Against this background of pervasive uncertainty, a few known quantities do emerge. Hogarth's first and, as it turns out, only contribution toward decoding the transmission is to demonstrate that it possesses a mathematically closed structure. It closes "like an object separated from the rest of the world, or like a circular process,"

[12]For an example of a puzzled response to this lack of unity, see Peter S. Beagle, 1983.

[13]Monod discusses teleonomy in an argument that sets forth in biological terms the same problems that Hogarth faces in shaping his text (1972:3–44).

Hogarth says, then adds, "to be more precise, like the DESCRIPTION, the MODEL of such a thing" (6:86). It is possible to hear in Hogarth's language a nostalgic yearning for a lost world of unproblematic referentiality. By establishing that the letter possesses formal properties of closure, Hogarth has indeed limited what it can be. He succeeds in proving that it is a "description of a phenomenon" (6:88). But when the research teams assign stereochemical meanings to the transmission's symbols, the text is inevitably opened again to chance.

From these assigned meanings the scientists are able to construct a physical substance. Whether this substance is what the letter was meant to describe remains problematic, however, since there is no way to be certain that the assigned meanings correlate with the letter's symbols. The substance may be a correct materialization, or a garbled distortion that reflects only the anthropomorphic biases of its creators. The ambiguity is reinforced by the uncertain purposes to which the substance might be put. Is it a miracle compound, revealing secrets hitherto unknown to terrestrial science? Or is it a first step in the Senders' plan to take over the planet? The ambiguity is encoded in the names the scientists choose for the substance. The biophysicists call it "frog eggs"; the biologists, "Lord of the Flies." Whether the substance is a tool for creation or destruction is as uncertain as whether it is the product of chance or design.

A third known quantity emerges from the letter's form, when it is discovered that the radiation composing the transmission has precisely the right range of frequencies to affect favorably the development of large protein molecules. If it had been present when Earth was young, it would have significantly increased the chances that life would evolve. It is, in other words, biophilic. Arguing from premises familiar to literary critics, Hogarth believes that since the *form* of the letter is conducive to life, the *content* must also be benevolent. He therefore pooh-poohs the notion that the letter may be intended for evil purposes, that it may, for example, describe a life form that will take over the planet as soon as it has materialized. In the absence of any context for the letter, however, Hogarth admits that his belief in the essential goodness of the Senders is faith, not fact. Although there is thus something that can be said about the letter, some facts and tangible products have been derived from it, its meaning remains ambiguous.

From this uncertainty comes Hogarth's severest test of faith. The occasion arises when Hogarth's friend and colleague Donald Prothero thinks of subjecting "frog eggs" to the same neutrino wave of which it is a materialization. Thus the transmission is turned back on itself, in a circular dialectic reminiscent of the circles-within-circles that are recurrent structures in the novel. This reflexivity suggests that the experiment will be crucial, and it is. The irradiation of "frog eggs" with the neutrino wave unleashes an awesome energy that, if projected on a suitably large scale, could turn into a doomsday weapon.

With this discovery, the ethical ambiguities implicit in Hogarth's account reach their height. A peace advocate, Hogarth has been suspicious from the start of the Pentagon's role in the project. His suspicion has been reinforced by the physical restrictions placed on the project's personnel. When Hogarth joined the project, he perforce also entered the secret desert compound that housed it. The psychological distaste he feels for the project's secrecy is manifest in his claustrophobic physical response to the enclosed compound. But now he and Prothero find themselves compelled to form a conspiratorial circle within this larger circle of secrecy, for they know that it is only a matter of time before someone else also discovers that the substance can be used to create explosions. As the circles-within-circles tighten into increasingly oppressive enclosures, the conspirators are put in the position of doing exactly what they most fear others will do—developing the weapon. Their panic about the implications of what they have discovered makes them desperate to find some way to gain control over the situation. Hogarth fleetingly entertains the idea of declaring himself dictator of the planet, using the weapon to *force* peace on humanity. He quickly realizes, however, that such a paradoxical venture would lead only to further oppression and to an even more claustrophobic enclosure.

The essence of the trap within which Hogarth finds himself caught is a context so coercive that it completely determines action, regardless of individual motives or beliefs. Hogarth has been darkly imagining that the Pentagon is funding the project because it hopes that a weapon will emerge. There is a fine moment when someone suggests to him that the Pentagon is involved because it is *afraid* a weapon may emerge and cannot take the chance that it may not be under its control. Thus the Pentagon, like Hogarth and Prothero, is shown to be the victim of a coercive context that forces it toward global annihilation, regardless of what its leaders may or may not want to do.

From such subtleties, the picture slowly emerges of an entire society caught up in a coercive context that determines how its texts will be read. When Prothero writes to government officials insisting that information about the project should be shared with the Russians and proposing that the two nations should work together to crack the code, his letter serves only to identify him as an object of suspicion. The militarists refuse to consider his plan because they are convinced that the Russians would only *pretend* to cooperate, meanwhile setting up a second, real team in secret. Prothero is outraged by this answer. Its probability is indirectly confirmed, however, when Hogarth discovers that this is exactly what the Pentagon has done with them. He learns that there is not one project, but two. "His Master's Voice" is shadowed by a second, even more secret project, appropriately named "His Master's Ghost" (13:153). The doubling, with its suggestion of self-reflexivity, emphasizes that the human text performs as if it were void, emptied of content, when it is enclosed within a sufficiently coercive context.

For Hogarth, being trapped within a context that seems inevitably to be leading him and everyone else toward global destruction is the nadir of experience. It is the nightmare of his mother's death, returned and magnified. In a long, drunken wake for humanity, he and a confidante try to imagine what accidents of fate, what chance events were fossilized into the history of the race to cause the present context to emerge. In his antibiography, Hogarth first saw chance as a way to fight evil, then concluded that chance and necessity were inextricable. At this moment, chance becomes for Hogarth evil as well as necessary. The dialectic has become a vicious circle.

The way up and out is characteristically a further inversion of, rather than a retreat from, this conclusion. Unable to escape its coercive context, humanity is delivered by what Hogarth interprets as a triumph of text over context. When the future seems darkest, Prothero discovers that the explosion's accuracy diminishes with range, so that it is worthless as a weapon. Hogarth comes to believe that only the mistranslations of earth's "predatory" science made the weapon even seem possible. The weapon fails to work, he believes, because the Senders were able to create a tamperproof text, a message so immune to misinterpretation that not even the coercive context of global war could empty it of its benevolent content. Just as the threat of the doomsday weapon enacted on a magnified scale Hogarth's first personal encounter with death, so the Senders' text achieves on a cosmic scale the solution Hogarth sought when he dedicated his life to the study of chance. The Senders deliver humanity from impending annihilation because they have mastered chance.

Are we to endorse this conclusion? As Hogarth admits, it is not necessarily true. Other explanations are possible. It remains for Hogarth an act of faith, a belief that ethics can be recuperated because chance can be controlled. But then Hogarth places the control of chance at the center of his own life as well, so his conclusions about the neutrino transmission are suspiciously parallel to his conclusions about his life. It is clear that for Lem, the answer is more complex than Hogarth, in his optimistic conclusion, wants to believe, for the ambiguities that Hogarth overcomes by an act of faith are reinforced as the novel draws to a close.

The final pages recount what happens when scientists from "His Master's Ghost" arrive to present their interpretations of the neutrino transmission. Whereas researchers from "His Master's Voice" have been assuming that the neutrino wave is a letter that implies Senders, those from "His Master's Ghost" have been working on the assumption that the transmission is a message that has form but no content, or a message that has no sender. These alternative possibilities undermine the reader's faith, if not Hogarth's, that humanity is saved because the Senders can control chance. There may be a voice, or there may be only a ghost. Ethics may emerge from the void, or it may not. Moreover, the reflexive mirroring between each project's name and its conclusions (the Voice project believes there is a sender, the Ghost project that there is not) reinforces the suspicion that each project reads itself in the message rather than decodes whatever is actually there.

These self-reflexive mirrorings and the uncertainties they engender are necessary to make sure that the textual space can never be definitively closed. For if the Senders *were* able to control chance, might not the cosmos itself someday become an oppressive enclosure? Only because the transmission remains enigmatic can the universe be open to both chaos and order, chance and necessity. Only because a hole gapes within the center of the text can the necessity to cease writing be enfolded together with the ongoingness of creation. For Lem, the encounter with chaos is less a celebration than an ambiguous necessity.

The dialectical enfolding of chaos into order serves both as the mark of Lem's affinity with his time and as the stamp of his individuality. In his concern with chaos and with textual dissemination, Lem has strong affinities with the new paradigms. Yet in his insistence on rational consistency and rigorous extrapolation from known facts, he can also be seen as very much at odds with the postmodern temper. To place him at either of these extremes is in an important sense not to see him at all, for it is from the tension between these polarities that his writing emerges. He is on the cusp of the transition from the old to the new paradigms but not actually within either, for he resists the synthesis the new paradigms offer at the same time that he exposes contradictions that the old did not recognize. His work testifies that during a paradigm shift, change arrives not by a single sweeping transformation but by complex compromises negotiated at specific sites.

Lem's contribution to the twentieth century's encounter with chaos is to reveal (perhaps inadvertently) an ambiguity within chaos theory which has not received the attention it deserves. As I remarked earlier, chaos theory has a double edge, at once celebrating chaos's resistance to rationalization and striving to overcome this resistance. Lem uses a similar opposition as the basis for his craft. In this respect he has fully come to terms with an ambiguity that most chaologists are only beginning to think about. His distinctive achievement is to use the tension between chaos and order to create a space for writing that cannot be enclosed even within its own self-erasing confines. We may trace where the dialectic has been by following the motion of the pen; but the dialectic itself is never written, always writing. . . .

Part II

THE FIGURE IN THE CARPET

CHAPTER 6

Strange Attractors: The Appeal of Chaos

IT is difficult to draw the line between what is genuinely new in the strange-attractor branch of chaos theory and what is the result of changed perceptions. Certain aspects of chaos theory were not previously known—that a meaningful distinction could be made between chaotic and disordered systems, for example, or that systems that appeared disordered could nevertheless manifest deep structures of order. From a scientific point of view, these discoveries perhaps justify calling chaos theory a new science (although, as we have seen, Poincaré's seminal papers date back to the 1890s). From a broader perspective, however, the thrust of chaos theory comes as much from the realization that nonlinearity is pervasive throughout the natural world as from new theories as such. Where the eighteenth century saw a clockwork mechanism and the nineteenth century an organic entity, the late twentieth century is likely to see a turbulent flow. The importance of chaos theory does not derive, then, solely from the new theories and techniques it offers. Rather, part of its importance comes from its re-visioning of the world as dynamic and nonlinear, yet predictable in its very unpredictability.

At issue in any re-visioning is a shift in what is taken as the norm. If chaotic systems exist now, they obviously existed before. Turbulence, for example, has been an object of investigation in fluid mechanics for virtually as long as the field existed. But only in the last few years has nonlinearity been recognized as representative rather than exceptional. Presumably observations confirming the extent of nonlinearity in the natural world could have been made at any time; the point is they were not. Similarly, scale dependence could have been observed in many phenomena before the emergence of the new paradigms. Although some examples were recognized, they were considered to be exceptions to the rule rather than to instantiate a rule. Also overlooked was the inability to specify initial conditions accurately enough to predict future states of the system. Although a wide variety of behaviors might have made this fact clear, the emphasis in Newtonian paradigms on long-term prediction directed attention away from the issue, making it seem like a nonissue. Joseph Ford, for example, points out that specifying initial conditions within physics has traditionally been regarded as "a task of such apparent triviality that no mention of it appears in the [technical] literature" (1983:44). The more subtle aspects of chaos theory have to do with changes in orientation and focus. For this reason, it may appear that the new science is simply discovering what everyone has known all along. There is some truth in this observation—but only some, for knowledge does not exist in an ideal space removed from culture. What is known is a function of what is noticed and considered important. If the criteria defining center and margin change, in a very real sense the structure of knowledge changes as well.

The power of a new paradigm to reinterpret well-known phenomena can be aptly illustrated through the swinging pendulum that is a textbook example of Newtonian mechanics. Suppose we construct a pendulum that consists of a ball at the end of a motorized shaft. Ordinarily the ball swings parallel to the driving momentum of the motor. However, if the frequency of the motor shaft's oscillations is close to the pendulum's natural period, the ball also develops a perpendicular motion. The interaction of this motion with the driving force causes the pendulum to become extremely sensitive to small variations in the starting conditions. In fact, a pendulum oscillating in this fashion has such sensitive dependence on initial conditions that it is impossible to measure the starting point accurately enough to ensure that the pendulum will swing the same way twice. David Tritton, who uses this device to demonstrate chaos in action, comments that if "we knew *exactly* how the pendulum is moving at a given time, then we could predict its future motion exactly. But we never do know anything exactly—the slightest vibration in the drive or the slightest draught in the room prevents that" (1986:37).[1] Thus the system that was emblematic of Newtonian determinism for the eighteenth century, a swinging pendulum, can easily be made to act unpredictably.

If chaos is everywhere, even in the deceptive simplicity of a pendulum, why was it not noticed before? Scientific journals are full of articles attributing departures from expected orderliness to errors, faulty experiments, or erratic equipment. Researchers interested in chaos theory are returning to these "noisy" data and testing them for the characteristic patterns they have learned to recognize; in a significant number of cases, the patterns are there. It seems clear that they were not noticed before because no paradigm existed through which they could be understood. In *Chaos* (1987), James Gleick documents how researchers in various fields slowly became aware that others were working along similar lines and that what they were seeing were not aberrations but a new kind of normality. The emergence of chaos is virtually a casebook demonstration that scientific investigation is not simply a matter of objectively describing nature.

What is it a matter of? Chaos theory is still so new that little work has been done on how it became a new paradigm. To my knowledge there are only two explorations of how this paradigm shift came about. One is an appendix to Benoit Mandelbrot's *Fractal Geometry of Nature* (1983), in which he argues that the mathematics associated with chaos was the work of mavericks who refused to follow received ideas. The second is Gleick's *Chaos*. Gleick, a science writer for *The New York Times*, concurs with Mandelbrot's view that chaos emerged from the work of a few solitary individuals. "No committee of scientists pushed history into a new channel," Gleick writes, "—a handful of individuals did it, with individual perceptions and individual goals" (p. 182). I believe we should view such explanations skeptically, for they fit too neatly into the American

[1]Tritton's discussion is intended for a general audience. A more technical treatment is D. d'Humières, M. R. Beasley, B. A. Huberman, and A. Libchaber, "Chaotic States and Routes to Chaos in the Forced Pendulum," *Physical Review* A 26 (1982): 3483–3496.

myth of the rugged individual who single-handedly opens up a new frontier. I shall return to Gleick's book to discuss how it inscribes scientific investigation into narrative patterns that imply science is an individual enterprise rather than a reenactment of highly charged cultural concerns. At issue is not only what chaos implies as a new paradigm but what it can tell us about the culture in which it is embedded. First, however, it is necessary to understand more about chaos in its scientific sense.

The Deep Structure of Chaos

Before chaos became a recognized research field, Edward Lorenz happened upon its fundamental properties when he was trying to model the earth's weather. Lorenz thought that he could capture the essence of how weather changes through three nonlinear differential equations.[2] The nonlinearity was important, for it kept the patterns from repeating themselves exactly. However, Lorenz got more unpredictability than he bargained for. He had programmed his computer to display the solutions graphically, as wavy lines of symbols. Gleick tells the story of how Lorenz stopped a sequence in midcourse, and rather than start over from the beginning, typed the mid-point values into the computer to start the sequence running again (Gleick, 1987:15–18). Lorenz presumed that this mid-point run would match the first run exactly. To his surprise, when he compared the two lines he found that they diverged—at first only a little, but then more and more.

On reflection he realized that the display was amplifying a minute difference in initial conditions. The computer stored six digits in its

[2] A differential equation expresses the rate of change of one variable with respect to another; for example, how the pressure of a gas changes as the temperature is raised. If both variables change at the same rate (that is, if for every additional degree of temperature the pressure rises one unit), the graph expressing this relation will be a line tilted at a 45-degree angle between the two right-angle axes. If the relation between the variables is nonlinear, the graph will show a curve rather than a straight line. If it is strongly nonlinear, the curve will have a very sharp hump (or humps) in it. Only with linear differential equations can one be sure of arriving at a general solution. Some nonlinear differential equations can be solved, but these are special cases.

memory, but the printout displayed only three. When he started the mid-point run with these three digits, he was beginning from a very slightly different place than where the run had stopped. This small initial discrepancy was amplified until the final result was a pattern quite different from the original. Carried into weather forecasting, this result implied that a sea gull flapping its wings on a California beach could affect the occurrence of snow on Wall Street.

Fascinated, Lorenz began to explore the behavior of his model in more detail. He found that the data contained what would later be called a strange attractor. An attractor is simply any point within an orbit that seems to attract the system to it. A swinging pendulum, for example, eventually comes to rest at the mid-point of its period unless it is driven by a motor. Because this point is always the same, the pendulum is said to have a fixed-point attractor. Other oscillators have attractors that are cycles, such as the double rhythm of the human heart. When the heart is disturbed (provided the disruption is not too massive), it returns to this characteristic rhythm even if it begins from a different place. Because the cycle takes place within well-defined boundaries that limit the extent of variations within any given cycle, it is called a limit cycle.[3]

One way to model an attractor is to construct a phase space. Suppose we want to describe a swinging pendulum. For convenience we can choose one point in each swing and note what the pendulum is doing at that point. Say we choose the mid-point. Each time the pendulum passes the mid-point, we note its position and velocity. To see how these numbers change over time, we plot them on a graph, with the horizontal axis representing position and the vertical axis velocity (see figure 3).[4] Rather than show the pendulum's actual motion as a movie would, this graph corresponds to a series of snapshots depicting the pendulum's behavior at a single instant in each cycle. If the pendulum is driven by a motor at a constant rate, the

[3]For a discussion of heart rhythm as a limit cycle, see Ary L. Goldberger, Valmik Bhargava, and Bruce J. West, "Nonlinear Dynamics of the Heartbeat," *Physica* 17D (1985): 207–244. Arthur T. Winfree also has an excellent discussion in *When Time Breaks Down: The Three-Dimensional Dynamics of Electrochemical Waves and Cardiac Arrhythmias* (Princeton: Princeton University Press, 1987).

[4]Douglas Hofstadter has a clear and accessible explanation of phase space in "Metamagical Themas," *Scientific American* 244 (November 1981): 22–43.

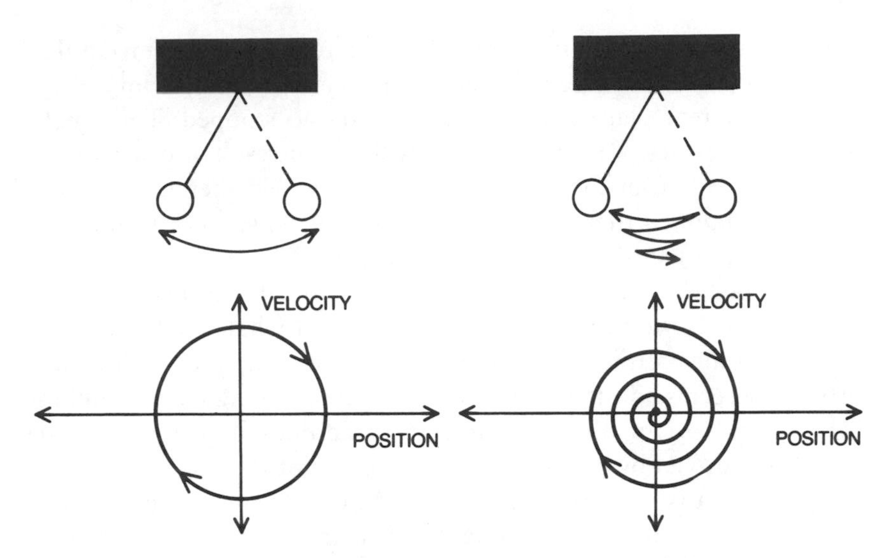

3. Phase space diagram of a pendulum being driven at a constant rate (*left*) and running down (*right*). Drawing by Andrew Christie from James P. Crutchfield, J. Doyne Farmer, Norman H. Packard, and Robert S. Shaw, "Chaos." Copyright © 1986 by Scientific American, Inc. All rights reserved.

graph will show a circle. If it is slowed down by friction, the diagram will show a spiral contracting toward the middle. Because there are two quantities that can vary—position and velocity—the system is said to possess two degrees of freedom, and the graph representing these variables is called a two-dimensional phase space.

It is important to remember that a phase space diagram is not the same as an ordinary drawing whose dimensions correspond to length, width, and depth. Phase space dimensions represent variables within the system, and phase space shapes show how they change over time. An orbit in phase space should thus not be confused with a trajectory through space. Rather, it is a map of the changes in the system's behavior over repeated cycles. Systems more complicated than pendulums will have more degrees of freedom and consequently will require greater dimensions for their representation. When the degrees of freedom exceed three, these systems cannot be easily drawn. Their mathematical properties, however, can still be explored.

Recall that Lorenz's model was represented by three nonlinear differential equations, corresponding to three degrees of freedom.

4. Phase space diagram of a Lorenz attractor. Courtesy of Herbert Hethcote.

When he graphed its motion in a three-dimensional phase space, he saw a shape resembling two butterfly wings, intricately traced by the orbits as they repeated the cyclic motion over and over (figure 4).[5] The orbits always stayed within a certain volume, so the shape was quite distinctive; but within this volume, no two orbits ever intersected or coincided, an indication that the system never repeated the same motion exactly. Because the orbits "wandered" in complex and unpredictable ways, knowing the starting point for a given run did not help one to know where the system would be at any future moment. Starting coordinates that were very close on the number scale could lead to orbits far apart; conversely, points numerically far apart might end in orbits close together. No wonder this attractor was considered strange. It was an odd combination of simplicity and complexity, determinism and unpredictability.

[5]See Lorenz, 1963, and his "Problem of Deducing the Climate from the Governing Equations," *Tellus* 16 (1964): 1–11.

What gives a phase space these peculiar properties? Drawing on early work by the distinguished mathematician Steven Smale, theorists realized that oscillators that behaved in this fashion acted as though the phase space within which they moved had been squeezed and then folded over, much as croissant dough is rolled out and folded over itself again and again.[6] This process creates an extremely complex interleaving of very thin layers. If it is carried on long enough, the layers approach infinite thinness, just as orbits in phase space have zero thickness. Nevertheless, each orbit remains distinct from every other; no two ever exactly coincide. James Crutchfield and his collaborators (1986) simulated this process through the computer manipulation of an image (they impishly chose a portrait of Henri Poincaré, the founder of dynamical systems theory). The simulation showed the outlines of Poincaré's face quickly becoming unrecognizable as the points spread out. But after a certain number of squeezings and foldings, the points aligned themselves so that the image briefly reappeared, as though the memory of the intact image lingered within the system (see figure 5). This odd combination of randomness and order conveys the flavor of a strange attractor.

The force of chaos theory derives from the discovery that an astonishing variety of systems can be modeled as strange attractors. For example, they have been shown to describe outbreaks of infectious diseases, variations in cotton prices and in the numbers of lynxes caught by trappers, the rise and fall of the Nile River, and erratic eye movements in schizophrenics.[7] Gleick reports that the group of graduate students at Santa Cruz who became pioneers in the chaos theory played a coffeehouse game of guessing where the nearest strange attractor was: in the din of dishes coming from the kitchen? in the swirl of cream in coffee? in the clouds of cigarette smoke coming from the next table (1987:201–226)? According to

[6]A discussion of this process is presented in Crutchfield et al., 1986. Steve Smale gives an account of his work in *The Mathematics of Time: Essays on Dynamical Systems, Economic Processes, and Related Topics* (New York: Springer, 1980).

[7]See, for example, William M. Schaffer and Mark Kot, "Nearly One-Dimensional Dynamics in an Epidemic," *Journal of Theoretical Biology* 112 (1985): 403–427; William M. Schaffer, "Stretching and Folding in Lynx Fur Returns: Evidence for a Strange Attractor in Nature?" *American Naturalist* 124 (1984): 798–820; Bernardo A. Huberman, "A Model for Dysfunctions in Smooth Pursuit Eye Movement," Xerox preprint, Palo Alto Research Center.

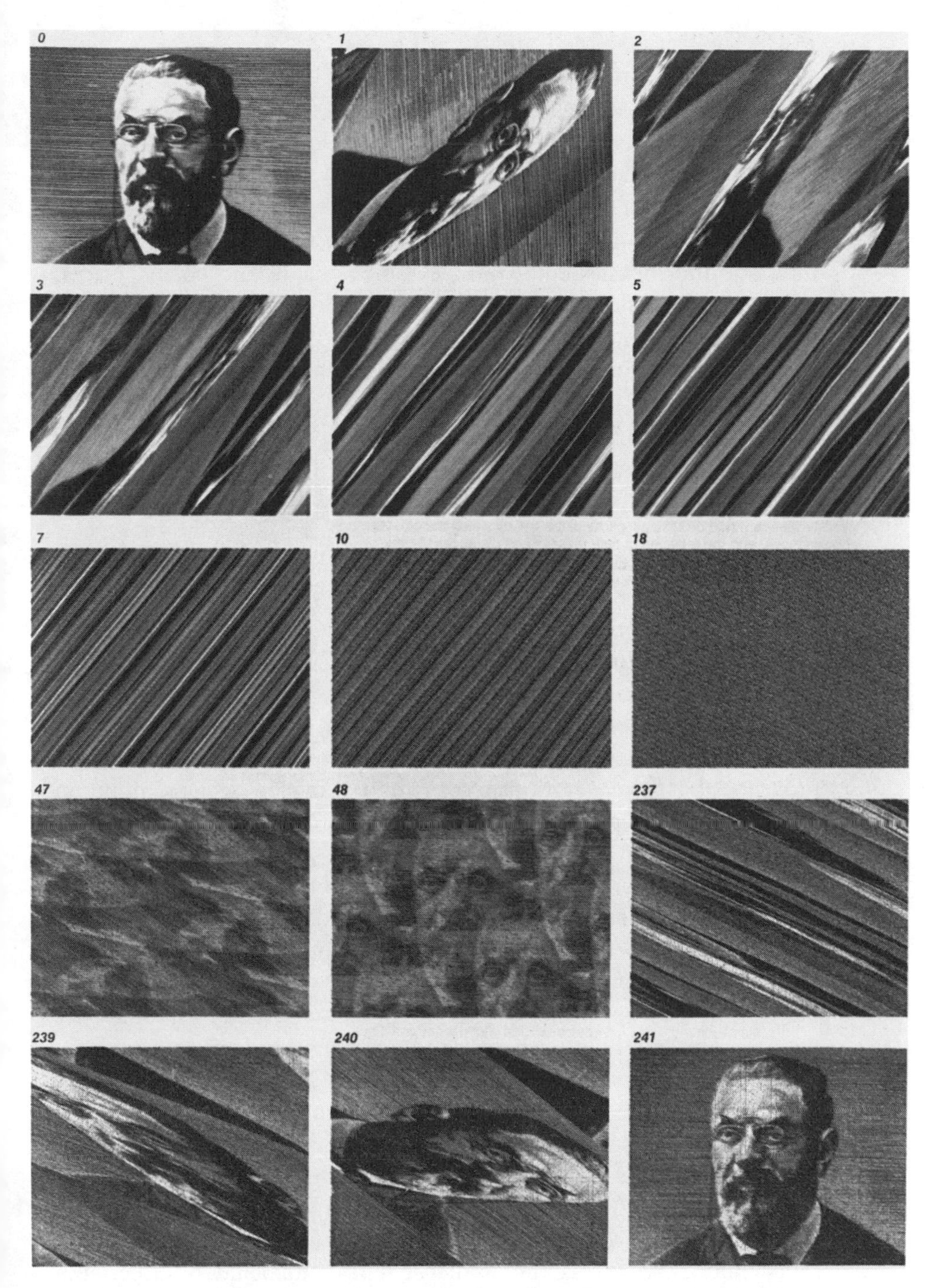

5. Stretching an image of Henri Poincaré. The number above each box indicates the number of times the stretching operation has been performed. Reproduced from James P. Crutchfield, J. Doyne Farmer, Norman H. Packard, and Robert S. Shaw, "Chaos," by permission of James P. Crutchfield. Copyright © 1986 by Scientific American, Inc. All rights reserved.

Gleick, a faculty member at Santa Cruz became convinced that his rattling speedometer was a strange attractor. Everywhere they have looked, chaologists have found seemingly erratic variations that prove to have the deep structure of chaos.

What does it mean when systems with so little in common as fur trapping and measles epidemics, eye movements and cotton prices, follow the same model? What has one understood when one creates a model that pays no attention to the specific mechanisms at work in the system? Do such models count as explanations? Do they even count as science? To begin exploring these questions, I turn to the work of Mitchell Feigenbaum, who perhaps more than anyone else deserves credit for recognizing the universal nature of chaos.

Chaos as Universality Theory

One approach to chaos is to consider the closely related concept of randomness. In a review article in *Los Alamos Science* (1980), Feigenbaum asks us to consider in what way a number generated by a computer can be random. He points out that a computer-based random number generator can easily be constructed by means of a program that does "nothing more than shift the decimal point in a rational number whose repeating block is suitably long" (p. 4). Numbers generated by this method are so lacking in pattern that they satisfy the most rigorous tests for randomness, yet the method that produces them is perfectly simple and deterministic. Technically, these numbers are called pseudo-random to indicate they have been generated by a deterministic computer program. Feigenbaum's inspiration was to wonder whether other apparently chaotic phenomena also might be pseudo-random, obeying deterministic programs just as pseudo-random numbers do. This astonishing premise amounts to saying that chaos has a deep structure.

One of the first indications a deep structure might in fact exist was Feigenbaum's discovery that systems that go from ordered to chaotic states follow a characteristic pattern of period doubling. Let us say that we are looking for a pattern in the behavior of an electrical oscillator. We notice the oscillator repeats its behavior after some time interval T. As the temperature is raised, the oscillator's behavior becomes more erratic, and we now have to extend the time period to $2T$ to have a repeating pattern. When the temperature is raised yet again, the time required to observe a repeating pattern jumps to $4T$, and so on. Eventually the time period will become so great that the oscillator has no repetitions in the time scale available for observation. At this point it is said to be chaotic. Period doubling is now recognized as a powerful generalization, describing the onset of chaos in everything from dripping faucets to Niagara Falls.[8] At the time of his discovery, however, Feigenbaum was not thinking about physical systems. He was looking at the behavior of mathematical functions when they are iterated.

To iterate a function means to use the output of one calculation as input for the next, each time performing the operation called for by the function. It is analogous to beginning at a certain place and doing a dance step; then starting from the new location each time, doing the dance step again and again. Iterating strongly nonlinear functions produces paths that have folds in them when they are displayed in phase space. The folded orbits Feigenbaum produced were simple one-dimensional maps. The behavior of these iterated functions, however, was analogous to the more complex stretching and folding of multidimensional phase space which is characteristic of strange attractors. "This general mechanism," Feigenbaum comments, "gives a system highly sensitive upon its initial conditions and a truly statistical character: since very small differences in initial conditions are magnified quickly, unless the initial conditions are known to *infinite precision*, all known knowledge is eroded rapidly to future ignorance" (p. 21).

The startling aspect of Feigenbaum's work was his discovery that despite the different operations performed by different nonlinear functions—despite the different dance steps they used—their iterated paths approached chaos at the same rate and showed the same characteristic patterns of period doubling. All that mattered was that the paths had folds of sufficient steepness. It may be hard for a nonscientist to appreciate the enormity of this discovery, but to a mathematician, sine waves are as different from quadratic equations as

[8]Robert Shaw, *The Dripping Faucet as a Model Chaotic System* (Santa Cruz: Aerial, 1984). Niagara Falls would of course be an instance of turbulent flow, as discussed in Wilson, 1983.

pirouettes are from bows to a ballet dancer. To find out that there is a way of looking at these functions that makes their operations seem not just similar but *identical* is analogous to discovering that there is a way of looking at Nureyev dancing and Donald Duck waddling that makes their performances into a universal constant applicable to anyone moving on that stage.

What was this new way of looking? At the same time that the iterative process had the effect of overwhelming individual differences between functions, it also revealed a universality in the way large-scale features related to small details. The shift in focus was from the particularities of a given function to the relation between different recursive levels in the iterative process. Imagine two paintings, each showing an open door through which is revealed another open door, through which is another and another. . . . One way to think about the doors in these two paintings is to focus on the particularities of the repeated forms. Suppose the doors of the first painting are ornately carved rectangles, whereas the second painting shows doors that are unadorned arches. If we attend only to shapes, the paintings may seem very different. But suppose we focus instead on the recursive repetition and discover that in both paintings, the doors become smaller at a constant rate. Through this shift in focus we have found a way of looking at the paintings that reveals their similarity to each other and to any other painting constructed in this way. The key is recursive symmetry.

Chaos and Symmetry

The mathematics Feigenbaum used to reveal recursive symmetries was developed by Kenneth Wilson, winner of the 1982 Nobel Prize in physics. To understand Wilson's approach, consider what happens when a flow becomes turbulent (Wilson, 1983). Often microscopic fluctuations within a flowing liquid cancel each other out, as when a river flows smoothly between its banks. In this case each water molecule follows much the same path as the one before it, so that molecules starting close together continue to be close. Sometimes, however, microscopic fluctuations persist and are magnified up to the macroscopic level, causing eddies and backwaters to form.

Then molecules that began close together may quickly separate, and molecules that were far apart may come close together. As a result it becomes extremely difficult to calculate how the flow will evolve. "Theorists have difficulties with these problems," Wilson explains, "because they involve very many coupled degrees of freedom. It takes many variables to characterize a turbulent flow or the state of a fluid near the critical point" (p. 583). Indeed, the mathematics of turbulence is so complex that even the new supercomputers are inadequate to handle it. Computation times become unreasonably long after only three or four variables are considered, whereas dozens are necessary to create a model that can simulate turbulence accurately. Hence the importance of being able to solve these problems analytically.

Wilson's contribution was to devise a method that would sometimes yield analytical solutions. The essence of his approach was to shift from following individual molecules to looking for symmetries between different scales. In turbulent flow, for example, large swirls of water have smaller swirls within which are still smaller swirls. . . . To model these recursive symmetries, Wilson used renormalization groups. Renormalization had first emerged as a technique in quantum mechanics; physicists used it to get rid of infinite quantities when they appeared in the equations. Originally the only justification for the procedure was that it made the answers come out right. If you think this sounds suspiciously aribitrary, you are not alone. For years virtually all mathematicians and some physicists regarded renormalization as no more than hand-waving. But Wilson saw its deeper implications.

He knew that in renormalization certain quantities regarded as fixed, such as particle mass, are treated as if they were variable. He realized that there is a sense in which this is a profound truth rather than an arbitrary procedure. For example, we tend to think of a golf ball as a smooth sphere. But to a mosquito it would appear as a pocked irregular surface, and to a bacterium as the Wilson Alps. Renormalization implied that the choice of ruler used to measure physical properties affected the answer. At the same time, it revealed that there was something else—something not normally considered—that remained constant over many measurement scales. This was the scaling factor. By combining the renormalization process with the idea of a mathematical group, Wilson arrived at a method whereby this factor could be defined and calculated.

"Group," as it is used in mathematics, denotes a set of objects that is invariant under symmetry operations. For example, a cube rotated 90 degrees in any direction appears unchanged in its spatial orientation; it is therefore said to be invariant under right-angle rotation. Tetrahedrons are in the same mathematical group as cubes because they also have this property. The purpose of finding a renormalization group is to look for symmetries that are invariant for different measurement scales. As an illustration of one of these symmetries, consider the classic "middle third" set, first proposed by Georg Cantor in the nineteenth century. Imagine that we draw a line from 0 to 1, as in figure 6.[9] Now we erase the middle third of line *a*. Each of the smaller lines of *b* has the same form as line *a*, and multiplying one of these line segments by 3 gives the original line back again. If we erase the middle third of the small lines of *b*, we create the even smaller lines in *c*. Each of the two broken lines of *c* has the same form as the entire interval of *b*, and multiplying either segment by 3 gives line *b* back again. If we keep erasing middle thirds, each time the symmetry of the resulting small part mirrors the larger part of the step before, and each time we can obtain the larger part by multiplying the smaller part by 3. Sets that have this kind of symmetry are said to possess fixed points. The purpose of defining a renormalization group is to discover the operations and variables that allow fixed-point symmetry to emerge.

Groups that display fixed points have physical significance because the symmetry allows coupling to take place between different levels. When a system possesses fixed points, perturbations on the smallest scale are quickly transmitted throughout the system, affecting even the largest macroscopic level. Imagine a bullwhip moving at just the right frequency so that a small twitch of the handle is transmitted into larger and larger waves all the way to the end, causing the whip to emit a loud, satisfying CRACK. This kind of transmission and magnification is possible because a system possesses an appropriate kind of symmetry. When fixed-point symmetry is present, systems "crack" because the symmetry permits microscopic changes to translate into coordinated movements all through the system.

[9]Cantor's "middle third" set is discussed at length in Mandelbrot, 1983:74–96.

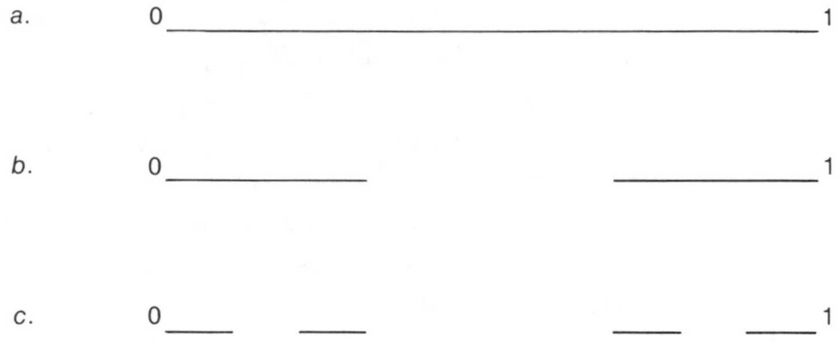

a. 0________________________________1

b. 0__________   __________1

c. 0___ ___   ___ ___1

6. Cantor's "middle third" set

We can appreciate the generality of this approach by looking at the range of behavior to which it applies. As we have seen, one area is turbulent flow; another is phase change behavior, as when water turns to steam. The transition from water to steam illustrates how fluctuations at the microscopic level can be translated into dramatic macroscopic changes. Water and steam appear to be very different—one is liquid, the other gas. However, this gross change in macroscopic properties originates in microscopic changes. At the point where the phase change takes place, the system is characterized by "bubbles of steam and drops of water intermixed at all size scales from macroscopic, visible sizes down to atomic scales" (Wilson, 1983:583). At this critical point, and only at this point, the system undergoes a phase change because it possesses the fixed-point symmetry that allows changes at the smallest level to be transmitted all the way up to the largest level. Thus turbulent flow and phase change behavior, different as they are, can both be understood in terms of the scaling symmetries. Additional applications are in quantum field theory, where coupling mechanisms between particles are important.

To see how this approach can be applied to chaos, consider it as an explanation for strange attractors. A strange attractor possesses fixed-point symmetry, implying it has very many coupling points that transmit and magnify tiny fluctuations into large changes within its orbit. The physicist M. V. Berry has likened this situation to playing tennis in an orchard.[10] Two successive balls may be hit in very nearly identical ways, yet, they will wind up in widely separated locations within the orchard because each time they are deflected by

[10]M. V. Berry, "Regular and Irregular Motion," in Jorna, 1978:16–147.

a tree, their trajectory changes. The more trees there are for the balls to hit, the more impossible it is to predict where they will end up. In this analogy, the trees correspond to coupling points, and the final location of the balls to points along a strange attractor's orbit after several iterations. We may know fairly precisely in what direction a ball is first hit; that is, we know its initial conditions. But unless we know *exactly* what the direction is—unless, as Feigenbaum pointed out, we know the initial conditions with *infinite precision*—we will not be able to hit a second ball so that it falls even close to the first. After a relatively short time, balls will be scattered throughout the orchard at locations that cannot be calculated in advance. Thus deterministic systems may also be chaotic.

One limitation of Wilson's method was its inability to tell whether a system possessed fixed points. If fixed points were there, it could yield an analytical solution; but it could not predict whether a solution was possible. Nevertheless, it was a breakthrough because it synthesized previous research into a single explanatory framework. And it gave Feigenbaum an approach to use in his efforts to understand chaos.

Whether Feigenbaum's work was a startling discovery or an extension of what was already known continues to be debated. Some scientists see Feigenbaum's universality theory as central to the science of chaos. Others regard it as derivative, especially since Feigenbaum's scenario of period doubling emerges as a special case of Wilson's group symmetries. Critics say that Feigenbaum was looking at only a narrow spectrum of chaotic behavior, and that work by Wilson and others was more powerful and general. Feigenbaum's concentration on iterative equations rather than physical models is also an issue; it was only when the same behavior was found in experimental data that chaos became a major research area. Nevertheless, Feigenbaum was the first to realize that scaling symmetries implied not just qualitative similarities but exact numerical correspondences. He was the first to see that chaos, in its way, had a structure as rigorous and compelling as order.

Chaos as Maximum Information

One of the curious chapters in the history of chaos is the role played by a group of graduate students at the University of California at Santa Cruz. Gleick reports how the group formed around Robert Shaw, constituted themselves as a collective, and began a research program that mingled investigations into chaotic systems with a fanatic interest in roulette (1987:201–226). As members of the collective tell it, they worked virtually alone; their advisers in the physics department neither approved nor understood the significance of what they were doing. Although each member made important contributions, Shaw is perhaps the best known. It was his idea to connect chaos with information.

In his article "Strange Attractors, Chaotic Behavior, and Information Flow" (1981), Shaw analyzes strange attractors by dividing their orbits into minimum blocks of phase space, with dimensions on the order of Planck's constant (the very small number that appears as a universal constant in the uncertainty principle). This minimum space is identified with a "logon" or information cell of a minimum size, thus translating the oscillator's orbit into a flow of information. Having the flow contract, as it does when an oscillator begins outside its limit cycle, is equivalent to saying that information once accessible has been destroyed. With strange attractors the reverse process takes place; information that was once inaccessible is created. Attractors operate irreversibly because their operation changes *what we can know about them*, not merely what we do know.

Shaw's analysis of an attractor's orbit as a flow of information creates a perspective fundamentally different from that of classical Newtonian dynamics. Unlike Newtonian paradigms, it does not assume that quantities are conserved. Information is not merely transformed, as matter and energy are. It really does come into being or disappear. Once an oscillator with a contracting information flow has made the transition to its steady state, the information contained in the initial conditions cannot be recovered. Conversely, knowing only the initial conditions and the relevant equations of motion will not allow one to predict where an oscillator with an expanding flow will be at some future time.

Since the equations governing a strange attractor's motion are deterministic, where does its unpredictability come from? Recall that Feigenbaum located the source of uncertainty in the initial conditions. He argued that this was the *only* place uncertainty could have entered, because the equations and the computer iteration were both deterministic. Shaw's focus on physical systems, by contrast, leads him to an experimentally based interpretation. He sees the chaos as coming from "hitherto unobservable information, such as a random fluctuation of the heat bath, [which] has been brought up to macroscopic expresssion" (p. 84). In this view the oscillator acts as a "translator" between the microscopic and macroscopic levels. Because the uncertainty principle guarantees that fluctuations always exist on the particle level, there will always be enough microscopic uncertainty to initiate macroscopic chaos. All that is required is for the system to have the appropriate symmetry so that it can act as a translator. Thus Shaw asserts that Newtonian determinism is "not only wrong in the small, where quantum mechanical uncertainty limits our knowledge of the exact state of the system, but it is wrong in the large" (p. 108).

The celebratory tone of his conclusion is strikingly different from the cautious sobriety found in most scientific journals. For Shaw chaos is the source of all new information in the world. He revels in the implication that chaotic systems are much more abundant in nature than ordered ones. Envisioning the physical world as a great swirling flow of information, he sees new information "continuously being injected . . . by every puff of wind and swirl of water." He acknowledges that this "constant injection of new information into the macroscales may place severe limits on our predictive ability." More important for him, however, it "insures the constant variety and richness of our experience" (p. 108). Shannon would have approved. Thirty years earlier he had intuited that chaos was information's partner. Even he, however, did not imagine that their coupling could bring forth the infinite fecundity of nature.

Chaos and the Limitations of the Human Mind

Despite Shaw's eloquent conclusion, not everyone sees infinite information as cause for celebration. Joseph Ford, a flamboyant physicist at Georgia Institute of Technology who runs an informal clearinghouse for chaos papers, argues that infinite information is essentially inhuman (1983). He believes that numbers harboring infinite complexity should be excluded from numerical representation because they exceed the ability of humans to understand them. His concern centers on results from algorithm complexity theory demonstrating that almost all real numbers are infinitely complex. When even such seemingly well-behaved creatures as real numbers are capable of chaotic behavior, they bewilder the humans who conceived them.

To understand Ford's concern, it is necessary to know something of algorithmic complexity theory. An algorithm is the set of procedures a computer uses to solve a problem, such as the instructions telling a computer to generate random numbers by moving the decimal point in a rational number. Algorithmic complexity theory is concerned with defining and calculating the amount of information an algorithm contains. The theory, invented independently by Gregory Chaitin (then an undergraduate at City University of New York) and A. N. Kolmogorov of the Academy of Sciences of the USSR, was proposed to solve the problem of how a number can be named (Chaitin, 1975).

The problem of how to name a number may seem trivial until one considers an example known as Berry's Paradox, first published in *Principia Mathematica* and reviewed by Martin Gardner (1979). Consider the number "one million, one hundred one thousand, one hundred twenty-one." This number appears to be named by the expression "the first number not namable in under ten words." However, this definition has only nine words. Is the number named by its nine-word definition, which says that the number cannot be named in under ten words? As Gardner points out, the paradox demonstrated that naming is too powerful a concept to be used without restriction (1979:20). To resolve it and similar paradoxes, algorithmic complexity theory restricts the concept of naming by saying that an integer has been named when it has been calculated by a computer program. Such a program then becomes the number's name. Algorithmic complexity theory defines the *complexity* of this name as the bit length (that is, the amount of information) of the shortest computer program that can print the given number sequence. Maximum complexity is achieved when the computer program generating the number is not appreciably shorter than the number itself.

To see how this idea relates to chaos, recall that, by definition, random numbers have no pattern to their sequences. Thus the algorithm generating a random number *n* might read, "Print *n*." But specifying *n* in this program takes about as much space as *n* itself, because the information contained in the number cannot be appreciably compressed. Thus the simplest way to specify the sequence is to provide a copy of it. Chaitin shows that almost all real numbers are random in the sense that they cannot be generated by a finite computer algorithm (1975:50). Following Mark Kac, Ford points out that "the [number] continuum therefore has the distinction of being a well-defined collection of mostly undefinable objects" (1983:44).

What does this discovery mean for human knowledge? Ford connects it with the importance chaos theory places on specifying initial conditions. As we know, unless the initial conditions are stated with infinite precision, the smallest uncertainty in a strange attractor can quickly translate into macroscopic chaos. Because most real numbers are random, however, they cannot be stated with complete accuracy. Thus unpredictability in deterministic systems is inevitable. In Ford's view, this realization will force a reformulation "whose impact on science may be as significant as [relativity theory and quantum mechanics]" (1983:46).

Ford's suggestions for the direction this reformulation might take are based on his belief that humans cannot cope with infinite complexity. He proposes that we move toward a "humanly meaningful" number system by eliminating numbers with infinite information content. First to go are the "totally inhuman numbers" that possess maximum complexity, numbers that "require infinite information to compute, to store and to define." Next are the "seemingly innocent" numbers computed by repeating a finite algorithm an infinite number of times. Last to be excluded are the infinitely large (1 + 1 + 1 . . .) and the infinitely small 1/(1 + 1 + 1 . . .). These successive eliminations would leave the number continuum so riddled with holes that Ford predicts all physical variables would henceforth be quantized (1983:47).

How far this view is from Shaw's celebration of chaos is apparent in the fable with which Ford ends his article. It is a mathematical version of the fall from Paradise. Man, at first content with simple integers, succumbs to temptation when he accepts an infinite number from his mate. The man's mind "reached and fleetingly grasped the meaning of (1 + 1 + 1 . . .), but by morning he retained only the empty symbols" (1983:47). Ford thus implies that from a human viewpoint, infinite information is indistinguishable from total incomprehensibility, a conclusion Borges anticipated in "The Library of Babel." One is accustomed to encountering such thoughts in *Ficciones*. What does it mean to read them in *Physics Today*?

The Beauty of Chaos

Many scientists working on chaos speak of the need to "develop intuition."[11] They point to the fact that most textbooks treat linear systems as if they were the norm in nature. Students consequently emerge from their training intuitively expecting that nature will follow linear paradigms. When it does not, they tend to see nonlinearity as scientifically aberrant and aesthetically ugly. But nonlinearity is everywhere in nature and consequently in mathematical models. Despite its prevalence, it has been ignored for good reason: except in a few special cases, nonlinear differential equations do not have explicit solutions.

A key factor in getting around this difficulty has been the development of microcomputers, which has led to a new style of mathematics. Ordinarily one does mathematics by stating a theorem and developing a proof. Mathematical reasoning thus proceeds through theorem–proof, theorem–proof, theorem–proof. But with computers, a new style of mathematics is possible. The operator does not need to know in advance how a mathematical function will behave when it is iterated. Rather, she can set the initial values and watch its behavior as iteration proceeds and phase space projections are displayed on a computer screen. Then she can see how the display modulates as she changes the parameters. The resulting dynamic interaction of operator, computer display, and mathematical functions is remarkably effective in developing a new kind of intuition. It is

[11]See, for example, Robert M. May and George F. Oster, "Bifurcations and Dynamic Complexity in Simple Ecological Models," *American Naturalist* 110 (1976): 573–599; and Benoit Mandelbrot, "Fractals and the Rebirth of Iteration Theory," in Richter and Peitgen, 1986:151–160.

perhaps the scientific equivalent of performance art. Whereas Ford saw the computer as purveying inhuman complexity, one could argue that computer displays of fractals and dynamic systems make complexity intuitively meaningful.[12]

The foremost advocate of a new *aesthetic* appreciation of nonlinear mathematics is Benoit Mandelbrot. In his quirky and fascinating book, *The Fractal Geometry of Nature* (1983), he argues that thinking that nonlinearities are strange is itself strange, since complex figures appear regularly in nature. He compiles adjectives that other mathematicians have used to describe nonlinear geometry—"monstrous," "counter-intuitive," "pathological," "psychopathic" —in much the same spirit as a Jesuit catalogues arguments refuting the existence of God, as an encyclopedia of misperception and error. Mandelbrot insists that on the contrary, highly complex and irregular forms are entirely compatible with our intuition and as beautiful as Nature herself. This beauty has been misperceived as "monstrous" because traditional geometry is ill equipped to deal with its complexities. "Many patterns in Nature," Mandelbrot writes, "are so irregular and fragmented, that, compared with *Euclid* . . . Nature exhibits not only a higher degree but an altogether different level of complexity" (p. 2).

To deal with the complexities of these forms, Mandelbrot invented fractal geometry. At its heart is the idea of a fractional dimension. Whereas Euclidean shapes can be represented very well by the familiar integer dimensions of a Cartesian space, irregular forms cannot. The corrugations that mark their surfaces give them, in effect, an added fraction of a dimension. Mandelbrot points out that the idea of fractional dimensions can also be applied to strange attractors. Earlier I compared the fixed-point symmetry of a strange attractor with a cracking bullwhip. Now I can refine that analogy by noting an important difference between an attractor and a whip. For the whip to crack, the length of the wave traveling down it must be such that an even number of wavelengths can fit into the total whip length. A strange attractor, by contrast, characteristically has an or-

[12]This aspect of computer technology is discussed in David Campbell, Jim Crutchfield, Doyne Farmer, and Erica Jen, "Experimental Mathematics: The Role of Computation in Nonlinear Science," *Communications for the Association of Computing Machinery* 28 (1985): 374–384.

bit in phase space which is a fractional dimension of that space. If an attractor moves in a two-dimensional phase space, for example, its orbit may have dimensions of two and two-thirds. Every known strange attractor has this property. Just as the corrugations on complex surfaces impart an added fraction of a dimension, so the extremely complex wanderings of strange attractors' orbits give them fractional dimensionality. The same principle applies to other complex trajectories. Planets that move in regular paths have orbits that are integers of phase space, for example, whereas irregularly moving bodies such as asteroids have orbits of fractional dimensions.[13]

Strange attractors and wandering asteroids are a sideline for Mandelbrot. His main interest is fractal geometry, which he more or less invented by exploring forgotten paths and labyrinthine turns in earlier mathematics. He coined the word "fractal" from the Latin adjective *fractus* (meaning "broken") and fractional; it connotes both fractional dimensions and extreme complexity of form. An important difference between fractal and Euclidean geometry is the scale-dependent symmetries of fractal forms. In Euclidean geometry, and in classical physics generally, Nature is considered "conformable to itself," of uniform consistency, so that what is true on one level is also true on another. In Euclidean geometry, for example, one equilateral triangle is taken to be similar to any other equilateral triangle, regardless of their relative sizes. For a geometry built on this assumption, the convoluted orbits of strange attractors look very strange, if not altogether deviant. But in fractal geometry the emphasis on recursive symmetries makes such orbits appear natural, because they are created by iteration of the same form over and over.

The complex couplings between scales of different lengths which are at the center of fractal geometry are found everywhere in nature—in cloud forms, mountain contours, tree grains. For these and many other natural forms, Mandelbrot explains, "the number of scales of length . . . is for all practical purposes infinite." Hence the importance of a geometry designed to elucidate transitions between levels by investigating and describing the symmetry operations that

[13]A related discussion is in Jack Wisdom, "Meteorites May Follow a Chaotic Path to Earth," *Nature* 315 (1985): 731–733.

make them possible. So complex are these forms that it is extremely difficult to duplicate them precisely, even by Mandelbrot's techniques. Thus one speaks of *approaching* the natural form, rather than representing it exactly. Because of this complexity, Mandelbrot stresses that "the most useful fractals [i.e., those found in nature] involve *chance* and both their regularities and their irregularities are statistical." Scaling, as Mandelbrot uses the term, does not imply that the form is the *same* for scales of different lengths, only that the degree of "irregularity and/or fragmentation is identical at all scales" (1983:2).

In his richly illustrated text, Mandelbrot encourages readers to see computer-generated fractal figures as natural contours, pointing out that one resembles a mountain range, another a seacoast, still another the branching streams of a watershed. The comparisons are not merely fanciful, for much of the usefulness of fractals lies in their ability to model such naturally complex forms as the human vascular system and the lumps and clusters of galactic star systems. One of the figures that Mandelbrot discusses is Cantor's "middle third" set, previously mentioned as an example of fixed-point symmetry. When the diagram given earlier in figure 6 is extended into increasingly fine divisions, the resulting figure, called "Cantor dust," can be used as a model for the distribution of galaxies in the universe. When the horizontal lines of figure 6, extended for several more levels, are made to flow vertically into one another to form two "draped" curtains, the figure models the breaks in Saturn's rings (1983:74–83).

Mandelbrot's procedure for generating fractals can be illustrated through the "Koch curve," named after its inventor, Helge von Koch. Imagine an equilateral triangle; suppose that we place smaller triangles, one-third the size of the original, on each side. We now have a hexagonal star, a Star of David. Suppose that along the sides of each point we place additional triangles reduced by one-third. Then on each of those triangles we place further triangles, again reduced by one-third, for as far as we wish to go (see figure 7). The increasingly complex curve traced by the boundary of this figure is obviously symmetrical, yet also possesses a complexity that approximates the irregular symmetries of a coastline, as can be seen in figure 8.

7. Generating a Koch curve. From Benoit B. Mandelbrot, *The Fractal Geometry of Nature* (New York, 1983), p. 47, by permission of W. H. Freeman Company.

The problem with such figures is that their complexities are too regular to fit most natural objects. We can improve the fit by "tampering" with the program to introduce a statistical element. Why introduce chance instead of generating more complexities deterministically? Because the objects to be modeled, such as coastlines, are themselves formed by chance. A deterministic approach "would be not only tedious, but doomed to failure," Mandelbrot argues, "because each coastline is molded through the ages by multiple influences that are not recorded and cannot be reconstituted in any detail. The goal of achieving a full description is hopeless, and should not even be entertained" (p. 210). Rather, he advocates injecting chance into the computer algorithms by using the theory of probability, the "only mathematical tool available to help map the unknown and the uncontrollable. The power of chance is widely underestimated," Mandelbrot asserts. "The physicists' concept of randomness is shaped by theories in which change is essential at the microscopic level, while at the macroscopic level it is insignificant. Quite to the contrary, in the case of the scaling random fractals that

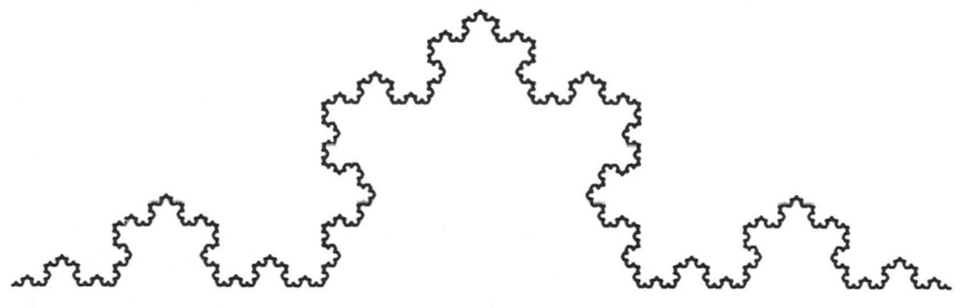

8. Making a Koch curve resemble a coastline. From Benoit B. Mandelbrot, *The Fractal Geometry of Nature* (New York, 1983), pp. 43–44, by permission of W. H. Freeman Company.

concern us, the importance of chance remains constant on all levels, including the macroscopic one" (p. 201). Hence Mandelbrot's final figures are combinations of symmetries generated by deterministic algorithms and injections of chance caused by stochastic perturbations of the original algorithm.

The story of how Mandelbrot became interested in fractal geometry has the quality of a parable. He insists it originated with a book review "retrieved from a 'pure' mathematician's wastebasket" (p. 422). His passion for resurrecting the forgotten and marginalized extends to people. In an appendix he gives brief biographies of earlier mathematicians who saw bits and pieces of what he would later make into a unified field. He sees them as avatars of himself, men who stepped out of their time to look at chaos in a new way. Their inclusion in his book fills a double role. On the one hand, they legitimate his enterprise by creating a tradition into which he can insert himself. On the other, their piecemeal accomplishments and peripheral status ensure that Mandelbrot can still claim for himself the honor of being the inventor of fractal geometry. The two editions of *Fractal Geometry* reveal the same kind of double gesture. In the first edition, when fractal geometry was a new idea, Mandelbrot took pains to present it as a continuation and extension of earlier ideas. By the second edition, when fractal geometry had become a recognized field that was attracting more researchers daily, Mandelbrot rewrote key passages to emphasize the uniqueness and singularity of his vision. There is an obvious political agenda to this jostling for position; it is no secret that Mandelbrot is a contender for a Nobel Prize.

Personal considerations notwithstanding, Mandelbrot's hunch that the most important contributions to chaos have been made by "mavericks" raises intriguing quesions about the relation of scientific content to style. Unlike many areas in the physical sciences, notably high-energy physics, chaos theory did not require expensive equipment operated by teams of researchers for its development. All it really required was an inclination to look at things from a different perspective. At least three of the field's superstars—Wilson, Feigenbaum, and Mandelbrot—see themselves as having made breakthroughs because they were interested in questions others had abandoned as unprofitable or suicidal. "The key seems to be time to spare," Mandelbrot writes; the sentiment is echoed by Feigenbaum and Wilson.

Probably most people in the physics community would agree that the style of chaologists is markedly different from the prevailing style in the community—more informal, eccentric, flamboyant. Is this style an outward and visible sign of a different way of thinking? And if so, is this new mode of thought merely the result of individual genius, as Mandelbrot implies, or it is in part culturally determined?

Chaos and Culture: Deep Assumptions of the New Paradigms

I remarked earlier on questions raised by models that do not take into account a system's specific mechanisms. Gleick reports that Bernardo Huberman met with incredulity when, at a conference on chaos, biology, and medicine, he demonstrated that a chaos model could account for erratic eye movements in schizophrenics (Gleick, 1987:227–230). A physiologist objected that Huberman had not taken into account the muscles responsible for moving the eye; a biologist pointed out that additional complexities were presented by the inertial effects of jelly within the eye. From these points of view, Huberman's model was a coincidence rather than an explanation. Huberman's critics thought that a proper explanation for erratic eye movements should show how the eye functioned as a unit and how it was connected to neural and muscular groups. The anecdote illustrates a point made by Thomas Kuhn (1970) in a different context: when a paradigm shift occurs, the very nature of what counts as an explanation changes. Explanations under two different paradigms are not just dissimilar; they are incommensurable.

In the chapters to come, I explore the nature of this paradigm shift and connect it with developments in critical theory, literature, technology, and culture. Its general outlines are already clear. Newtonian paradigms focus on individual particles or units. The assumption is that if these units are followed through time, their collective actions will add up to the system's behavior. The emphasis is therefore on isolating the appropriate unit and understanding the mechanisms that bind units into larger groups. It is a foundationalist approach, infused with assumptions about the integrity and autonomy of the individual.

The fundamental assumption of chaos theory, by contrast, is that the individual unit does not matter. What does matter are recursive symmetries between different levels of the system. Chaos theory looks for scaling factors and follows the behavior of the system as iterative formulae change incrementally. The regularities of the system emerge not from knowing about individual units but from understanding correspondences across scales of different lengths. It is a systemic approach, emphasizing overall symmetries and the complex interactions between microscale and macroscale levels. From this perspective, a proper explanation is one that is able to model large-scale changes through the incremental evolution of a few iterative equations. Hence the emphasis in Huberman's model was not on the eye as such but on the equations that, when iterated, would produce wandering orbits equivalent to the eye's movements. From his point of view it was coincidental that the system in question was an eye. It could equally well have been fibrillation in the heart or fluctuations in cotton prices.

When one thinks about the great paradigm shifts of the past, one tends to suppose that the new paradigm was so obviously superior to the old that people were more or less instantly persuaded of its value. I suspect, however, that this was almost never the case. For people at the time, a new paradigm seemed to occlude as much as it revealed. It is only in retrospect that the occlusion seems insignificant and the revelation momentous, for once we are on the other side of the divide, we cease to care about the older issues. Today many scientists feel that chaos theory occludes essential aspects of a system's behavior; others believe that its revelations are immensely significant. It is difficult to adjudicate between these views, because they involve fundamental cultural assumptions that extend beyond the scope of scientific theory. Whether, for example, it is more important to decide if angels have enough corporeality to limit the number that can dance on the head of a pin, or to be able to describe mathematically how a cannonball's trajectory changes through time, depends on a multitude of cultural factors and not just on the merits of the theories alone. Scientific change *is* culturally conditioned.

To illustrate how arguments for the autonomy of science are culturally based, I should like to turn to a text I have referred to throughout this study, Gleick's *Chaos*. Although *Chaos* is written to be accessible to a general audience, it is no facile popularization. Meticulously researched, it draws from hundreds of personal interviews and cites many of the major technical articles in the field. It is also beautifully written, combining a journalist's eye for human interest with a scientist's interest in concepts. I recommend it to anyone interested in the subject. I certainly learned a great deal from it, and I am grateful to the author for making it available to me before its publication.

Gleick describes his book as a narrative history. He rightly sees it as telling a story, and like any storyteller, he has shaped his material in ways both obvious and subtle. It is this shaping that I want to explore, for it depicts the discovery of chaos as essentially an individual enterprise, unaffected by the culture around it. I shall begin my story about Gleick's story at the point where it began for me—when, on a second reading, I realized that this book has no women in it. Hundreds of men are mentioned by name; some dozen are depicted in enough detail so that one almost feels one knows them. But no women, or virtually none, appear.[14] (Another reading revealed that women do appear a few times in the text, but only as anonymous "wives" who accompany their scientist-husbands at social events. In one case, Gleick implies that their presence transformed the meeting of two scientists into a social occasion, making it impossible for the men to conduct serious scientific conversation [p. 141].) I had a divided response to this absence of women. On the one hand, Gleick can scarcely invent them where they do not exist. Perhaps even more than most sciences, chaos is heavily dominated by men, especially in America. On the other hand, the exclusion of women in Gleick's text goes beyond the acknowledged scarcity of

[14]Several women are listed in the references. They include those whose names appear as co-authors of scientific papers; Erica Jen as a respondent (260); and Mary Lucy Cartwright as a researcher in chaotic oscillators (272). Evelyn Fox Keller has a fine analysis of masculinist assumptions in science in *Reflections on Gender and Science* (New Haven: Yale University Press, 1985). See also Ruth Bleier, *Science and Gender: A Critique of Biology and Its Theories on Women* (Oxford and New York: Pergamon, 1984), and Sandra Harding, *The Science Question in Feminism* (Ithaca: Cornell University Press, 1986).

distinguished women scientists. It pervades the entire depicted world.

The absence of women is most striking in the personal vignettes that Gleick uses to punctuate his explanations of scientific concepts. These vignettes show what the living quarters of this or that scientist are like; they reveal where one scientist goes for walks and what another likes to eat. But in all this richly textured detail, women never appear. The impression is that none of these men has a relationship with a woman which is important in his intellectual life; none works with a female collaborator who is an important contributor; and none spends much time with women. Having noticed this, I also began to notice the settings in which the scientists appeared. Very often they emphasize the solitary nature of their intellectual lives. Feigenbaum is shown walking away from a group so he can observe a waterfall; on another occasion, a crucial insight comes when he walks away from a group outdoors and notices that their talk is reduced to babble by the distance. Another scientist likes to take long walks in the desert, sometimes leaving his family for weeks at a time. When living quarters are discussed, they are depicted as eccentric or antidomestic. Feigenbaum has no furniture; the Santa Cruz collective lives in a house littered with bean bag chairs. Food is equally rudimentary or odd.

No doubt these details are accurate. But by mentioning them and not others, Gleick creates an ambiance for the scientific discoveries he describes. As it is shown here, the world of science is first of all genderless—genderless because there is only one gender. It is also solitary, with chance connections made between individuals who discover, often quite by accident, that someone else somewhere in the world is working on the same problem they are. And it is marked by a flow of narrative time in which certain moments are retrospectively identified as decisive, even though they may have seemed ordinary enough when they occurred. Treating time in this way is effective in creating the kind of suspense that keeps readers turning pages; but it also makes time into a series of agons marking the junctures at which fate took a different turn. These components work together to substantiate Gleick's claim that "no mass of scientists" brought about a new paradigm, only "a few solitary individuals" (p. 152).

Once I noticed how the material was shaped, I began to wonder how it fitted into the larger view of chaos that Gleick presents. For many of the scientists whose words he records, chaos is more than just another theory. It represents an opening of the self to the messiness of life, to all the chaotic unpredictable phenomena that linear science taught these scientists to screen out. Once roused, they remember that the messiness was always there. Moreover, now they are able to see nonlinearity in a new light, perceiving it as central rather than marginal, beautiful rather than aberrant. Chaotic unpredictability and nonlinear thinking, however, are just the aspects of life that have tended to be culturally encoded as feminine. Indeed, chaos itself has often been depicted as female. In the English Renaissance, for example, the male seed was commonly represented as contributing form; the female was thought to contribute raw unshaped *materia*, matter devoid of form or structure. In validating chaos as a scientific concept, Gleick seems to have found it necessary to expunge the female from his world. Why?

I can of course only speculate about the psychological and cultural dynamics underlying this exclusion. Nevertheless, certain aspects are sufficiently clear as to be almost obvious. In the Western tradition, chaos has played the role of the other—the unrepresented, the unarticulated, the unformed, the unthought. In identifying with chaos, the scientists that Gleick writes about open themselves to this otherness, and they perceive their intercourse with it as immensely fructifying—for their work, for their disciplines, and for them personally. But otherness is also always a threat, arousing the desire to control it, or even more extremely to subsume it within the known boundaries of the self, thus annihilating the very foreignness that makes it dangerously attractive. Both of these impulses are evident in Gleick's text, and probably are at work within chaos theory as well. The desire to control chaos is evident in the search for ways to rationalize it. By finding within it structures of order, these scientists have in effect subsumed chaos in the familiar. But if this incorporation were entirely successful, chaos could no longer function in its liberating role as a representation of the other. Perhaps this is why Mandelbrot goes out of his way to argue against the complete rationalization of chaos; he believes that some residue of the untamable and nonrational should always remain.

This ambivalence toward chaos is encoded into Gleick's text, especially in the divided response toward the feminine. Representations of actual women and of activities closely associated with them are rigorously excluded from the depicted world. Paradoxically, this exclusion facilitates the incorporation of the feminine principle of chaos into science. By admitting the feminine as an abstract principle but excluding actual women, Gleick attains control over the polysemy of chaos, striping it of its more dangerous and engendered aspects. As a result, chaos is admitted into the boundaries of scientific discourse, but science remains as monolithically masculine as ever.

In achieving this accommodation, Gleick's text engenders a series of paradoxes. It depicts chaos theory as the achievement of extraordinary individuals who stepped out of the mainstream; but this scenario of the solitary man who opens up a frontier is itself deeply a part of the American mainstream. It shows science as an exclusively male domain; but it is the peculiar project of this domain to have intercourse with a feminine principle. It intimates that scientific discovery is an activity that men engage in when they separate themselves from their families and from the larger culture; but the theories these men formulate imply that the individual unit is not important. The complex play of gender, individuality, and scientific theory in Gleick's text suggests that chaos theory is a deeply fissured site within the culture, in which lingering assumptions from older paradigms are embedded within the emerging paradigms of the new science. When there is this kind of complex interplay between science and culture, science cannot be separated from the cultural matrix. Like literature, science is always already cultural and cannot be otherwise.

I do not believe that the scientists Gleick writes about acted in isolation. I think that they rather acted like lightning rods in a thunderstorm or seed crystals in a supersaturated solution. They gave a local habitation and a name to what was in the air. It was because the cultural atmosphere surrounding them was supercharged that these ideas seemed so pressing and important. Mandelbrot, though he clearly wants to claim the territory for himself, recognizes that earlier mathematicians had penetrated his domain. Because the time was not right, their forays were regarded as dead ends. Now, for reasons that are as complex as the chaotic systems the new paradigms represent, the time is right.

CHAPTER 7

Chaos and Poststructuralism

At the same time that new concepts of chaos and randomness are changing the way scientists think about informational systems, they are also affecting the way literary critics write about texts. The major impetus for this revision has come from poststructuralism, especially deconstruction. Paul de Man, for example, sees deconstruction as warning us that "nothing, whether deed, word, thought or text, ever happens in relation, positive or negative, to anything that precedes, follows or exists elsewhere, but only as a random event whose power . . . is due to the randomness of its occurrence" (1979a: 69). As the text is opened to an infinitude of readings and as meaning becomes indeterminate or disappears altogether, chaos apparently reigns supreme. In this extreme form, deconstruction seems to have gone beyond the premises that make science possible.

Yet Geoffrey Hartman (1976), confronted with the "tangled, contaminated, displaced, deceptive" text of Derrida's *Glas*, speculates that dcconstruction is opposed to more traditional, humanistic readings because it is more *scientific*. "The result for our time [of deconstruction in general and *Glas* in particular] may be a factional split between simplifying types of reading that call themselves humanistic and indefinitizing kinds that call themselves scientific," Hartman writes (p. 183). He is correct, perhaps in a sense he did not intend, in linking deconstruction's "indefinitizing" strategies with science.

Deconstruction shares with chaos theory the desire to breach the boundaries of classical systems by opening them to a new kind of analysis in which information is created rather than conserved. Delighting in the increased complexity that results from this "scientific" process, both discourses invert traditional priorities: chaos is deemed to be more fecund than order, uncertainty is privileged above predictability, and fragmentation is seen as the reality that arbitrary definitions of closure would deny.

In this chapter I draw parallels between poststructuralist philosophy and scientific attitudes toward chaos, then discuss particular interpretive strategies employed by Jacques Derrida, Roland Barthes, and Michel Serres. In these analyses poststructuralism is my subject, not my method. Whereas poststructuralist analyses commonly operate upon texts to open within them fissures, discrepancies, and inversions, my analyses operate upon poststructuralist texts to show that they share with modern science assumptions and methodologies that can hardly be explained without the assumption that both are part of a common episteme. My approach is constructive rather than deconstructive. It seeks to delineate an ecology of ideas, to see similarities between scientific and literary theories as interrelated propositions that appear in separate discourses concurrently because they are responses compatible with the cultural environment.[1] This approach prevails especially in the section that takes Derrida and nonlinear dynamics as its focus.

An ecological approach is not, however, able to explain the very real differences between scientific models and critical theory. To elucidate how different evaluations can emerge from similar premises, I turn to an economic model. Information theory and poststructuralism concur in assigning a positive value to chaos. But where scientists see chaos as the source of order, poststructuralists appropriate it to subvert order. Thus the scientific response appears fundamentally conservative, whereas the poststructuralist stance styles itself radical. When institutional practices within the two disciplines are taken into account, however, both responses appear equally conservative, serving to perpetuate rather than challenge the economic in-

[1]The term "ecology of ideas" alludes to Gregory Bateson's *Steps to an Ecology of Mind* (San Francisco: Chandler, 1972). I am indebted to Bateson's illuminating analyses of communication dialectics throughout this chapter.

frastructures of their disciplines. Representative texts discussed in this section are Barthes's *S/Z* and Shannon and Weaver's *Mathematical Theory of Communication*.

The final section explores the synthesis of science and literature in the work of Michel Serres. A key concept for him is equivocation, the term that Shannon used to describe what is added to or subtracted from information as it passes through a noisy channel. Equivocation's inherent ambiguity—does noise add to information or subtract from it?—is present on a deep level in Serres's project. Demonstrating a strong proclivity toward order at the same time that it celebrates disorder, his work simultaneously tries to liberate noise at specific sites and to suppress it in global theories. The multiple levels at which these equivocations work within his discourse suggest that a theorist who locates himself at the crossroads of disciplines is able to maintain this position because he employs linguistic and conceptual structures capable of mediating between different disciplinary economies.

Throughout the three sections, I am concerned with rifts as well as convergences, differences, and similarities. Yet the final impulse is to show that both scientific and literary discourses are being distinctively shaped by a reevaluation of chaos. It is this vision that defines the contemporary episteme and differentiates it from the modernist era.

Ecology

As is well known, deconstruction grew out of developments in linguistics, particularly the Saussurean distinction between the signifier and the signified.[2] As long as this distinction is accepted as valid, we remain in the realm of semiotics rather than deconstruction. As Saussure recognized, however, the signified is a concept rather than an object. It is a small step to consider the signified as another signifier in turn; this step carries us out of semiotics into deconstruction. For if the signified is another signifier, then its signified is also a

[2]For an overview see Jonathan Culler, *Ferdinand de Saussure* (New York: Penguin, 1977) and *The Pursuit of Signs: Semiotics, Literature, Deconstruction* (Ithaca: Cornell University Press, 1981).

signifier, and we are caught in an endless chain of signifiers whose meanings, if they can be said to exist, are indeterminate.

Saussure initiated semiotics by proposing that the proper study for linguistics was *la langue* rather than *la parole*, the system of language rather than words, and by showing that relations within *la langue* could be specified only as a series of differences. The analogous movement within science came with Shannon's theory of information. Saussure's theory separated sign from referent; Shannon's theory separated information from meaning. In Shannon's equations, the informational probability of an element can be calculated only with reference to the ensemble from which it is drawn, that is, not absolutely but through a series of differences.[3] This move allows the information content of a message to be quantified *regardless of its context or meaning*. The inward-turning structures of Saussurean linguistics and information theory are not arbitrary theoretical choices. In the absence of external reference, these theories could be defined *only* internally. (I explore the cultural implications of these isomorphic strategies in chapter 10, arguing that they both reflect and reinscribe the postmodern condition.)

With deconstruction, the deeper implications of defining language and texts through their internal relations became explicit. Derrida's *Of Grammatology* (1967; English translation 1976) announced the birth of this deconstructive "monstrosity" (1976: 5). Until now, Derrida says, Western thought has persisted in believing in a Logos capable of revealing immediate truth. Speech is privileged because it is the embodiment of Logos, the word that is also presence. Writing, by contrast, is belated, a fall from presence into absence, the signifier of a signifier, inferior because twice mediated. These assumptions are inverted in *Of Grammatology*, which proclaims the beginning of a "grammatological" epoch in which writing is privileged over speech.

At the center of grammatology is Derrida's redefinition of writing. In the grammatological view, writing ("écriture") is not merely written marks but any signifying practice that endures through time and functions to divide the world into self and other. This view of writ-

[3]For a clear explanation of how difference enters into Shannon's theory, see Norman Abramson, *Information Theory and Coding* (New York: McGraw-Hill, 1963), pp. 1–44.

ing was apparently inspired by Freud's short essay "The Mystic Writing Pad," in which he likened the psyche to the child's toy that in this country is known as a magic slate.[4] The conscious mind, Freud suggested, is like the transculent sheet on which the child marks with a pointed stick. The marks are visible because the sheet is momentarily pressed into a wax base underneath. When the sheet is raised, the marks disappear. But they have not entirely vanished; their impressions linger in the wax base. The earliest experiences of childhood, Freud argued, are like the marks on the sheet, disappearing from the conscious mind as the child matures. But they leave traces in the deeper strata of the psyche, just as the marks leave traces in the wax. For Derrida, writing includes not only marks on the page but these deeper traces within the psyche. Residing at a deeper level than words can reach, the Derridean trace remains inaccessible to direct verbalization. It is "always already" present, the elusive and ineffable difference from which all subsequent inscription derives. In this sense writing, as the mark of difference, not only precedes speech but actually brings it into being.

By making the trace inaccessible and indeed unknowable, Derrida opens writing to radical indeterminacy. In contrast to speech, writing operates according not just to Saussurean difference but to Derridean "différ*a*nce," a neologism that in French combines "to differ" with "to defer" (Derrida, 1968). The "always already" formula implies that there is no origin, that the very idea of origin is an illusion. Différance acknowledges a before and after—that is, a constituting difference—but defers indefinitely the moment when this split occurred. No matter how far back we go into signification, we never come upon the originary difference that could act as ground for the chain of signifiers.

After writing *Of Grammatology*, Derrida moved on to other metaphors to uncover fissures within Western metaphysics—the pharmakon, the postcard, the hymen—presumably to avoid having his project reify into a methodology that could become a metaphysics in turn. Although there are suggestive parallels with the new paradigms

[4]My discussion of the trace is conjectural, as it is important to Derrida not to define it. Derrida discusses Freud's essay and its implications for deconstruction in "Freud and the Scene of Writing," in *Writing and Difference*, trans. Alan Bass, pp. 196–231 (Chicago: University of Chicago Press, 1978).

in much of this later work as well, the resonances are perhaps strongest in the early work in *Of Grammatology*, which also corresponds most closely in time to the first wave of important discoveries in chaos theory. I therefore will concentrate on this work, recognizing that it is only one phase of Derrida's continuing development of deconstructive methodologies.

That there *is* a methodology at work is a disputed claim, for positing a method is a first step toward establishing an orthodoxy that could quickly defeat the intent to unravel and undermine. Derrida himself is extremely canny about recognizing and avoiding this danger; even as early as *Of Grammatology*, he warned that his readings were too "exorbitant" and "excessive" to fit into any standard methodology (1976: 157–164). Nevertheless, one can scarcely fail to notice that deconstructive analyses tend to follow a predictable pattern, especially in the hands of disciples less adroit than Derrida in avoiding the dangers of reification. The method appears in paradigmatic form in *Of Grammatology*. First a received duality (speech/writing) is destabilized by inversion; then the existence of a third term (the trace) is revealed whose nature is undecidable because by definition it falls outside the realm of discourse. The validity of the speech/writing duality is thereby drawn into question, but no new dialectic emerges to take its place because the emergence of a dialectic would depend upon being able to decide what the trace is. In the grammatological approach, the original concepts do not entirely disappear but are put under erasure ("sous rature"). Seen and not seen, absent and yet present, they function as reminders that the old meanings are gone and as remainders that keep new ones from forming. Thus Derrida does not intend simply to replace one set of priorities with another. Rather he attempts to undermine the very process by which meaning is constituted.

The vertigo characteristic of deconstruction appears when we realize that texts are always already open to infinite dissemination. Far from being ordered sets of words bounded by book covers, they are reservoirs of chaos. Derrida initiates us into this moment in *Of Grammatology* through his concept of iteration. Any word, he argues, acquires a slightly different meaning each time it appears in a new context. Moreover, the boundary between text and context is not fixed. Infinite contexts invade and permeate the text, regardless of chronology or authorial intention. *Hamlet*, for example, influences our reading of *Rosencrantz and Guildenstern Are Dead*; but *Rosencrantz and Guildenstern Are Dead* also influences our reading of *Hamlet*. The permeation of any text by an indefinite and potentially infinite number of other texts implies that meaning is always already indeterminate. Because all texts are necessarily constructed through iteration (that is, through the incremental repetition of words in slightly displaced contexts), indeterminacy inheres in writing's very essence.[5]

We can see iteration at work in the dense, highly repetitive analysis of Rousseau that occupies the last half of the *Grammatology*. Rousseau is well suited to Derrida's deconstructive project because his thought is expressed through a series of hierarchical dualities: nature/culture, animal/human, speech/writing. For Rousseau, the first term of each of these dualities is privileged. The second term is belated, contaminated, a fall from the "pure" first term. His announced aim is to correct modern decadence by returning to the originary first term, rejecting culture for nature, writing for speech, and so forth.

Through a rigorous reading of Rousseau's texts, Derrida shows that this attempt at purification is fundamentally misguided because the idea of origin is an illusion. The demonstration concentrates on the "supplément," a word that Rousseau uses in the *Confessions* as a euphemism for masturbation. Sex is natural, good, healthy; but tormented by fear of women and venereal disease, Rousseau continually finds it necessary to resort to the supplement. Derrida shows that a similar dialectic emerges with each set of Rousseau's dualities (1976: 141–164). Rousseau denounces writing but does so by writing texts; he embraces nature but finds that its deficiencies require the education he advocates in *Emile*, and so forth. To supplement something implies that the original is already full and self-sufficient,

[5]For a discussion of iteration, see Derrida, *Of Grammatology* (1976), pp. 157–162. The concept of iterability also plays a prominent role in "Signature Event Context" (1977b) and in Derrida's subsequent deconstruction of John Searle's "Reply" to "Signature Event Context" in "Limited Inc abc . . ." (1977a). A fuller analysis of how iteration enters into the Derrida–Searle exchange is given in N. Katherine Hayles, "Disciplinarity and the University: What It Keeps Us from Seeing," in *The Nature of the University*, ed. Peter Shane (Iowa City: University of Iowa Press, forthcoming).

in contrast to the supplementary material, which comes after and is superfluous. Yet in each case the first term—"nature," for example—is "naturally" deficient, so that the supplement is indispensible. In what sense then is the supplement more "unnatural" than nature? Through this implicit contradiction, Derrida shows that the supplement is in fact what *allows the privileged term to be constituted.* The originary precedence of the privileged term is revealed as an illusion, a myth or longing for origin rather than an origin as such.

According to Derrida, every text will have a concept that functions as the supplement does in Rousseau. The supplement (or its analogue) is, Derrida argues, a kind of fold in the text whose indeterminacy is revealed through repetition.[6] In his view such a fold is *necessarily* present, because there must always be some means by which the text can constitute the differences that enable it to postulate meaning. The fold can be thought of as a way to create the illusion of origin. Once it is in place, all subsequent differences are declared to derive from the originary difference marked by the fold. When the text is "unfolded," this stratagem is revealed and the regulated exchanges between the alleged origin and subsequent differences that enable the text to operate will appear.

It is precisely this "unfolding" that iteration accomplishes. In Derrida's hands, repeating Rousseau's language with incremental differences becomes a way to unfold and make visible the inherent contradictions upon which the text's dialectic is based. This iterative procedure produces the undecidables that radically destabilize meaning. "It [is] certainly a production," Derrida writes, "because I do not simply repeat what Rousseau thought of[the supplement]. The concept of the supplement is a sort of blind spot in Rousseau's text. . . . [The production of undecidables] is contained in the transformation of the language [that the text] designates, in the regulated exchanges between Rousseau and history. We know that these exchanges only take place by way of the language and the text" (1976: 163–164). The goal of iteration is thus to make visible the lack of ground for the alleged originary difference, thus rendering all subsequent distinctions indeterminate.

[6]Derrida speaks of a "certain deconstruction which is also a traced path" (1976:162).

Derrida's deconstructive methodology is strikingly similar to the mathematical techniques of chaos theory. Recall that Feigenbaum attributed the universal element in chaotic systems to the fact that they were generated from iterative functions (1980: 4–27). He showed that for certain functions, individual differences in the equations are overwhelmed as iteration proceeds, so that even though the systems become chaotic, they do so in predictable or *regulated* ways. Derrida claims that his iterative methodology is similarly regulated, in the sense that its production of undecidables is not a capricious exercise but a rigorous exposition of the text's inherent indeterminacies.

For both Derrida and Feigenbaum, iterative methodology is closely tied in with the concept of the fold. Feigenbaum showed that systems that make orderly transitions to chaos always have folds in their iterative paths (p. 9). Within the complex regions created by these folds, orbits wander in unpredictable ways. Where does this unpredictability come from? Since the iterative formulae and computer algorithms are perfectly deterministic, it could come *only* from the initial conditions. Iteration produces chaos because it magnifies and brings into view these initial uncertainties. Similarly, Derrida attributes textual indeterminacy to the inherent inability of linguistic systems to create an origin. In Derrida, "always already" marks the absence of an origin, just as inability to specify initial conditions with infinite accuracy does for Feigenbaum. Thus nonlinear dynamics and deconstruction share not just a general attitude toward chaos, but specific methodologies and assumptions.

There are, of course, also significant differences between them. Feigenbaum works with mathematical formulae that are capable of exact definition; Derrida is concerned with language, which is notoriously resistant to formalization. One measure of these differences is disagreement among deconstructionists and scientists on how extensive chaos is. For Derrida, textual chaos is always already in Rousseau and in every other text. Scientists, by contrast, acknowledge that ordered, predictable systems do exist, although they are not nearly so widespread as classical science had supposed. Feigenbaum, for example, takes for granted that only certain classes of iterative functions become chaotic. Moreover, he acknowledges that until very recently, virtually all scientific knowledge derived from the study of ordered systems (pp. 14–15). Whereas *Of Grammatology* forecasts an apocalyptic break with logocentrism (a position that Derrida was to modify and complicate in later work), scientists are likely to think of their work as a continuation of what has gone before. To a deconstructionist, a "recuperator" is beyond salvation; for most scientists recuperation is not even an issue, because they see their work as enhancing rather than discrediting traditional scientific paradigms. Gregory Chaitin is typical in his view that algorithmic complexity theory merely *supplements* classical probability theory rather than supplants it (1975: 48). To the Derrida writing in *Of Grammatology* such a statement would itself invite deconstruction, for if the theory is supplemental it cannot be necessary, but since it *is* necessary, it cannot be supplemental. . . . These differences are symptomatic of the different values the two camps place on chaos. For deconstructionists, chaos repudiates order; for scientists, chaos makes order possible.

These differences notwithstanding, Derridean deconstruction and nonlinear dynamics are strikingly parallel in a number of ways. They agree that bounded, deterministic systems can nevertheless be chaotic; they both employ iteration and emphasize folds; and they concur that originary or initial conditions cannot be specified exactly. Given Derrida's antipathy toward science (for him "objective" is a pejorative term), it is unlikely that his ideas are substantially indebted to scientific sources. It is equally difficult to believe that Feigenbaum, Ford, or Shaw has been influenced by Derrida. How to account for the similarities?

The two theories appear isomorphic not because they are derived from a common source or because they influenced each other, but because their central ideas form an interconnected network, each part of which leads to every other part. Feigenbaum worked on iterative functions in part because advances in computer technology made it easy to see how they performed over time. He was led to the idea of an unreliable origin not because he was interested in the questions of origin as such, but because it was the only place chaos could have entered the system. For Derrida, working in a field dominated by Hegel, Nietzsche, and Heidegger, the question of origin was highly charged. Once he postulated a lack of originary ground, iteration was an appropriate methodology because it has the effect of magnifying latent uncertainties. Thus Derrida and Feigenbaum entered the network at different points and for different reasons. But because chaos, iteration, and an unreliable origin form an interconnected system of ideas, the correlative concepts were brought into play once the implications of the original premise were explored. Deconstruction and nonlinear dynamics appear isomorphic, then, because the concepts with which they are concerned form an ecology of ideas.

A theoretical model for conceptual ecologies has been proposed by Stanislaw Lem (1981), who suggests that they can be modeled as closed topological spaces. Within a given topology, only certain forms are possible. Others are prohibited by the overall spatial conformation.[7] Not every possible form will be realized; particularities of history and personality determine which actually appear and which are repressed. All forms that are realized, however, are linked to each other through the common attributes that define the space.

Lem concedes that so many variables are at work within contemporary culture that it can never be rigorously modeled as a topological space. Topological models have, however, been successfully used in biology to predict (or, more accurately, to retrodict) what appendages will appear for a given environment and set of genomes. These programs are limited to retrodiction rather than prediction, for in a highly complex system it is not possible to predict exactly how the system will evolve because there are too many cusps where minute fluctuations cause wide variations in future behavior.[8] Once one knows what has happened, however, it is possible to show that the realized forms are consistent with the system's topology. I conjecture that this is the case for nonlinear dynamics and deconstruction. They were not inevitable developments, but they were among the possible shapes that could evolve within postmodern culture. They are similar because they share in the constraints that define the overall topology.

[7]For an explanation of topology, see Bert Mendelson, *Introduction to Topology* (Boston: Allyn & Bacon, 1962).

[8]Catastrophe theory comes to very similar conclusions. See René Thom, *Structural Stability and Morphogenesis: An Outline of a General Theory of Models*, trans. D. H. Fowler (Reading: W. A. Benjamin, 1975). For discussions of catastrophe theory in relation to biology and linguistics, see René Thom, *Modèles mathématiques de la morphogenèse* (Paris: Union Général d'Editions, 1975).

An advantage of the topological model is that it leaves room for other kinds of explanation. It designates which shapes are possible, but it does not say which will actually appear, or how their actualizations are affected by such factors as disciplinary traditions and individual desires. To understand why deconstruction and the science of chaos value disorder differently even though they model it in conceptually similar ways, we will need to consider what happens when ideas are expressed within the boundaries of specific disciplines and articulated through specific voices. Why, for example, does deconstruction want to break with the past, whereas nonlinear dynamics wants to emphasize historical continuity? How do isomorphic concepts change when they are expressed in the tropes and linguistic registers peculiar to an individual theorist? To explore these and other questions, we turn from ecology to economy and equivocation.

Economy

Warren Weaver remarks that Shannon's theory of information is powerful because of its economy of explanation (Shannon and Weaver, 1949: 114–115). Implicit in the comment is the assumption that the best theory is that which can explain the most diverse phenomena with the fewest principles. The tendency in science is to simplify, to reduce the many to the few—millions of chemicals to some hundred elements, for example, then a hundred elements to three atomic particles. When the atomic triad proliferated into hundreds of subatomic particles with the advent of high-energy physics, the scientific community was disturbed; economy had been violated. An intensive search was undertaken for a grand unified field theory.[9] When its outlines began to emerge, its power was understood to reside in its ability to reduce an unruly mob to an elegantly simple number again, the four forces that govern the organization of matter. Shannon's theory of information falls squarely within this scientific tradition. His theorems are powerful in the sense that they reduce many different kinds of message signals to a few limit cases.

The passion for economical expression seems to have been a per-

[9]An overview of grand unified field theory can be found in Davies, 1984.

sonal as well as a professional aesthetic for Shannon. Colleagues recall him as a brilliant theorist who could not bear to relinquish his ideas for publication until they were expressed in the most economical form imaginable. In an anecdote recounted in *Grammatical Man* (Campbell, 1982), Edward Moore recalls that Shannon "would let a piece of work sit for five years, thinking it needed to be improved, wondering if he had made the right choice of variable in this or that equation. Then, while he was still contemplating improvements, someone else would come out with a similar result that was correct, but so lacking in formal elegance that Shannon would have been ashamed to have done such a shoddy job."

In contrast to Shannon's dedication to economy is Roland Barthes's exuberant expansiveness in *S/Z*. *S/Z* has a certain structuralist orientation in its impluse to classify and categorize Balzac's story *Sarrasine* in accordance with five discrete linguistic codes. Significantly, however, it uses these codes not to reduce but to expand Balzac's text. Thus it is even more poststructuralist, sharing with deconstruction the desire to increase message ambiguity as much as possible—for example, through split writing, dense syntax, elusive wordplay, and elliptical style. Barthes's project is ideological as well as aesthetic, for like Derrida he regards "correctness" as an illusion perpetrated by centrist philosophy to control texts, language, and power structures within society. In this respect information theory and *S/Z* could scarcely be more different.

At first glance, it appears that the political stances implied by these different aesthetics are easily defined. Shannon's theory is conservative because it seeks to wrest order out of chaos, preserve message from noise, and guard correctness from contamination by error. *S/Z* is radical because it seeks to liberate noise from message, release chaos from order, and overturn the hegemony of received interpretations. But because these theories do not act in isolation—because they are embedded within the economic infrastructures of their disciplines—their actual effects are more difficult to assess. I shall argue that when disciplinary contexts are taken into account, both theories are conservative, serving to perpetuate rather than challenge the traditions from which they come. First, however, it is necessary to look more closely at the transformations Barthes effects with Balzac's short story *Sarrasine* as he re-presents it in *S/Z*.

The radical stance that differentiates *S/Z* from information theory is apparent throughout Barthes's text. Declaring that he is not interested in what *Sarrasine* means, Barthes distinguishes five codes at work within his "tutor text" and identifies them with so many "voices" speaking *Sarrasine*. He refuses to arrange them hierarchically in search of a total meaning. Moreover, he points out that even within one code disparate connotations are often at work, as if two voices were speaking at once over the same channel. These "equivocations," as Barthes calls them, are to be encouraged:

> In relation to an ideally pure message (as in mathematics), the division of reception constitutes a "noise," it makes communication obscure, fallacious, hazardous: uncertain. Yet this noise, this uncertainty are emitted by the discourse with a view toward communication: they are given to the reader so that he may feed on them: what the reader reads is a countercommunication. [Barthes, 1974: 145]

Barthes concludes that "literatures are in fact arts of 'noise'" and declares that this "defect in communication" is "what the reader consumes."

The equivocation in Barthes's own text comes into focus with the word "consumes." With his assertion that literature is noise, he situates his project in opposition to the information theory from which he takes his terms. "Consumes" is a pivotal word because it recalls the capitalistic context of Shannon's career. Working at Bell Laboratories, Shannon was necessarily concerned with commercial applications of his work. He devoted considerable attention to how to get a message through correctly (Shannon and Weaver, 1949: 3–93). As we saw in chapter 2, Shannon defined any deviation from the intended message as the equivocation. One of his most important theorems demonstrates that it is possible to reduce the equivocation to zero if the code is chosen correctly.

The equivocation that Shannon wants to eliminate, Barthes offers up for consumption. Moreover, he claims that the reader will find this extra information more delectable than the original message. The persistent use of oral imagery in Barthes's text creates a new context for consumption, associating it with gourmandizing rather than capitalism. In effect, Barthes replaces the commercial orientation that sees correctness as an important value with a sensuality that finds its most exquisite satisfaction in deviant orality.

Within his discipline, however, Barthes's attitudes are not deviant; they merely express mainstream beliefs in a risqué fashion. For the economy of explanation that scientists regard with respect has long been viewed with suspicion in literary circles. Some critical methodologies have attempted a scientific economy of explanation—archetypal criticism and structuralism, for example—but they are the exceptions rather than the rule. In general the literary community favors convoluted explanations that expand the few to the many rather than economical explanations that reduce the many to the few. The phenomenon can, I believe, be understood in terms of the economic infrastructure of the discipline.

Before poststructuralism, literary criticism was confined to an accepted corpus of literary texts for its subject material. This body of texts remained essentially constant for decades, except for the influx provided by living writers; meanwhile, the academic literary establishment increased enormously.[10] Even when the influx of new writers is taken into account, the domain of critical theory has been extremely restricted in comparison with the scientific domain, where advances in technology and instrumentation continually open new areas for research. Who among us has not known the fear of arriving too late, when the texts have all been used up? Perhaps the Bloomian concept of belatedness is not so much analysis as projection, for even more than literature, literary criticism has operated according to an economy of scarcity. Too many critics, too few texts.

Poststructuralism, especially deconstruction, overcomes this scarcity by showing how each text can be made into an infinite number of texts. Moreover, it actually converts scarcity to excess by proclaiming that theory's proper subject is not only literature but theory itself.[11] In this sense, *S/Z* represents less the cusp between structuralism and poststructuralism than a harbinger and consort of decon-

[10]This argument neglects, for simplicity, the movement to open literary study to noncanonical works. As this trend can also be understood as a response to disciplinary economy, it would not change the essence of the argument, although it would render it more complex.

[11]For an appraisal and reluctant acquiescence in this claim, see Geoffrey H. Hartman, "Crossing Over: Literary Commentary as Literature," *Comparative Literature* 28 (Summer 1976): 257–276.

struction. By opening the text to dissemination and by blurring the distinction between "primary" and "secondary" works, deconstruction converts a closed system operating according to a scarcity economy into an open system based on autocatalysis. The more theory that is written, the more texts there are for theory to write about, because theory itself *produces* the texts that the next generation of theory will consume. As we saw in chapter 4, physical systems that are autocatalyzing can spontaneously reorganize themselves at a higher level of complexity (Nicolis and Prigogine, 1977). Theorists, unlike molecules, are conscious of the systemic organization they help to build. The analogy is useful, however, because it suggests that increased complexity arises not merely because of a decision by an individual theorist but because the systemic economy demands it. Thus the increasing number of theoretical texts in literary criticism, as well as their tendency to organize themselves in increasingly complex ways, can be understood as responses to the discipline's systemic economy.

At this point the reader may object that my explanation is "scientific," in the pejorative sense of reducing complex phenomena to simplistic generalizations, when it relates deconstruction's popularity to its economic function within the discipline rather than to its intellectual and conceptual power. I do not dispute that deconstruction is powerful. I would point out, however, that what it means for a theory to be powerful varies from discipline to discipline. We cannot properly evaluate the claim for deconstruction's power without first inquiring into the disciplinary economy that gives meaning to this term. Such an inquiry demonstrates that the aesthetics of deconstruction and information theory are consistent with their respective disciplinary economies. Moreover, their contrasting views of what counts as message and as noise are also consistent with these economies.

It is no accident, then, that conceptually isomorphic theories take on different values when they are embedded in disciplinary matrices. Meaning is always dependent on context, and different disciplinary economies create significantly different contexts. These contexts are reinscribed in Shannon's and Barthes's diverging views of equivocation. They agree that all messages are mediated. However, Shannon sees this as a regrettable fact of life, while Barthes envisions it as opportunity for repeated penetrations of the original text. They concur that noise is inevitable. But Shannon wants to minimize noise, Barthes to maximize it. Similarly, Shannon regards redundancy as a necessary evil, whereas Barthes sees it as an erotic pleasure that swells a compact text into a gargantuan commentary.

A discipline is not, then, only an abstract field of inquiry. It implies a specific workplace and set of institutional affliliations, and these in turn imply community norms that invest concepts with values. Shannon was an electrical engineer who worked for the world's largest information conglomerate. At AT&T, room in the channel translated directly into more expense for the company. Shannon's dedication to economy was reinforced by the multinational corporation that gave him a paycheck, as well as by the scientific community that gave him recognition. Consequently, he was not interested in just any economical explanation. He looked for one that would allow language itself to be more compressed. He did pioneering work on how to eliminate redundancy through proper coding (Shannon, 1951). His credo was compatible with the commercial economy within which he worked, as well as with the disciplinary economy that informed his aesthetic sensibility. Reduce and simplify, shorten and compress.

Barthes's approach is the opposite. Not only does he convert Balzac's 13000-word short story into a 75000-word analysis; he also implies that there is no valid way to compress this expansive interpretation. He defines the five codes in an appendix, for example, indicating that in compressed form they are peripheral to his enterprise rather than its concentrated essence. This ideology of excess is not, of course, unconnected to his economic situation. He writes as a critic within a literary establishment, where fame, money, and power come from generating new words from old texts. The more texts are opened to accommodate his words and others, the more the community in which he works will reward him.

The different values Barthes and Shannon assign to equivocation can now be brought into clearer focus. Recall that Shannon defined equivocation as information that the message sender did not intend. From a commercial point of view this information is superfluous, since the company transmitting the message is unlikely to be reimbursed for it. Consequently, Shannon assumed that the equivocation should be subtracted from the received message to get the original text back again. Barthes naturally disagrees, for if a text's meaning were limited to what the writer intended, the possibilities for literary criticism would be drastically curtailed. Since the canonized author is not paying for the message transmission, his intentions have no economic value. On the contrary, it is the consumers of Barthes's critical text who matter, and it is in their interest to see message noise increased, for then they are reassured that new and different books can be produced from the same canonized texts. Barthes therefore advocates that the equivocation should be added to the message rather than subtracted from it.

When Barthes contrasts rereading with consumption, he makes this difference in orientation explicit.

> Rereading, an operation contrary to the commercial and ideological habits of our society, which would have us "throw away" the story once it has been consumed (or "devoured"), so that we can then move on to another story, buy another book, and which is tolerated only in certain marginal categories of readers (children, old people, and professors), rereading is here suggested at the outset, for it alone saves the text from repetition (those who fail to reread are obliged to read the same story everwhere). . . . Rereading is no longer consumption, but play. [1974: 15–16]

The emphasis on play is thus Barthes's answer to Shannon's ideology of use. If texts are useful, then they can be used up. Only when they are infinitely equivocal, forever supplementing their original message with noise supplied by the reader, are they saved from the capitalistic economy that would consign them to obsolescence.

But wait. The situation cannot be this simple. Surely there must be room within disciplinary economies for moves that run counter to the mainstream or that deflect it in a different direction. Otherwise disciplines would be much more homogeneous than we know they are, and they would be static over time, which we know they are not. I turn now to discuss a countermove within information theory which foreshadowed the shift to chaos theory. I also interrogate Barthes's reading more closely, seeing in it a complex strategy having as much to do with his individual desires as with the disciplinary economy within which these desires are inscribed.

If Barthes had read Weaver's essay in *The Mathematical Theory of Communication*, he would have found the opposition sounding very much like him. Like Barthes, Weaver is a commentator. His essay originally appeared as an article in *Scientific American* and was intended to interpret Shannon's theory for a general scientific audience. Like Barthes, Weaver cannot resist expanding his role as commentator. He therefore speculates on how information theory might be extended beyond the strict engineering interpretation that Shannon gave it to include questions of semantics and behavior. As he leaves behind the terminology of use for the language of desire, his perspective undergoes a significant change. He suggests that into Shannon's diagram one might insert a box for "semantic noise," responsible for "perturbations or distortions of meaning which are not intended by the source but which inescapably affect the destination." This sounds conventional enough; however, he then goes on to say that "it is also possible to think of an adjustment of original message so that the sum of message meaning plus semantic noise is equal to the desired total message meaning for the destination" (Shannon and Weaver, 1949: 116). Thus the "desired meaning" goes from being what the sender intended to whatever comes out at the end after semantic noise has been included.

With this shift Weaver comes very close to Barthes's position. *S/Z* is designed to increase semantic noise as much as possible, both in the interpretations it brings to bear on Balzac and in the extensive commentary it physically inserts between Balzac's message units. When Weaver suggests that desired meanings could result from the *addition* of semantic noise, his formulation is in sympathy with Barthes's project to increase noise as much as possible. Is it a coincidence that a commentator should envision information not intended by the sender as being also desirable, perhaps even more desirable than the authorized message? However we wish to read Weaver's flirtation with equivocation, it is clear that here, in a text bound together with Shannon's original articles and read widely throughout the discipline, is a voice expressing views at odds with the disciplinary economy. Thus individual desire is also a factor affecting the way theories are interpreted.

For Barthes, desire and economy are mutually reinforcing. Yet economic motives that have their bases in disciplinary infrastructures may still be shaped in distinct ways by individual desire. To see how this works in *S/Z*, consider how Barthes chooses to decode his "tutor text." He writes as if the codes he distinguishes are preexistent in the culture, in the text, or in both together. As Shannon could tell him, this is not true. Codes are a matter of choice. They may be chosen well or ill, but they are always chosen, not inherent in the message.

The code that carries the most interpretive weight in Barthes's reading is what he calls the "Hermeneutic Code," the "Voice of Truth." This code traces Sarrasine's simultaneous pursuit and evasion of the truth that Barthes situates at the center of Balzac's tale: that Zambinella, the singer with whom Sarrasine has fallen violently in love, is not a woman but a castrated man. Barthes's decoding suggests that the exchanges in Balzac's story proceed according to an economy of castration. In his reading, all the characters are located in relation to the phallus—whether they want one, pretend to have one, or want to take one away from someone else. Moreover, castration is presented as being highly contagious, so that by the end of the story virtually every character is presented as having caught it, even those who never had a penis to begin with. Barthes thus reads the story as a superstructure erected on the economic base of the phallus.

A very different way to read the story is implicit in Barbara Johnson's critique of Barthes's interpretation (Johnson, 1980). She points out that Barthes *supplies* the word "castration"; Balzac's text never uses it. "Castration is what the story must, but cannot, say. But what Barthes does in his reading is to label those textual blanks 'taboo on the word castrato.' He fills in the textual blanks with a name. He erects castration into *the* meaning of the text, its ultimate signified" (p. 11). In contrast to Barthes's focus on castration, Johnson's reading puts narcissism at the center of Balzac's story. In her view, Sarrasine's transgression consists not in coming into contact with the highly contagious disease of castration but in preferring an essentially male fantasy of what a woman is—a man without a penis—to the radical otherness of woman herself.

Johnson's reading makes clear that Barthes's decoding of Balzac's text is not merely a neutral report on what is there but a strategy to gain control of the text's meaning. Disclaiming any desire to control the text, to silence or neglect any of its multivalent voices, Barthes nevertheless centers it on a void for which he then supplies the name. Moreover, his claim that it is *what has been added* to Balzac's text that most satisfies desire resembles Sarrasine's narcissism. Deleting from Sarrasine's actions the egotism of which he too is guilty, Barthes gets the pleasure of a narcissistic correspondence between his interpretive strategy and Sarrasine's character, while at the same time interpreting both so as to conceal this common feature.

Barthes cagily avoids the question why he choses *Sarrasine* as his "tutor text," but *Sarrasine* is ideal for his purposes despite its realistic texture, for it has an absent center in the missing phallus. Because this center is already empty, Barthes's strategy of filling it as he desires can be hidden from notice. Using the same strategy, Barthes attributes to information a centrality and to his "noise" a marginality that imply he does not share in the will toward mastery he attributes to information theory. But when noise occupies center stage—as it surely does in *S/Z*—this constituting difference between information and noise ceases to operate.[12] When the center is empty—or *has been emptied* by discourse that claims to be marginal—there is no meaningful distinction between margin and center in terms of the power they exercise. Thus Barthes's reading of Balzac cannot be understood solely through its reinforcement by a disciplinary economy. Also important are individual desires and agendas.

Like ecology, then, economy is an incomplete explanation. Ecology limits what can be but does not determine what is; economy influences what will be heard and attended to but does not adequately explain why a message is presented in a particular way in a given text. Since equivocation has emerged as a pivotal term, it obviously signals an intersection of some importance. In the work of Michel Serres, equivocation appears in highly fissured and self-re-

[12]The more general point is that deconstruction depends upon a rhetoric of marginality, which can be effective only when it is in a minority position. To the extent that deconstruction has become an accepted ideology within literary studies (if not, indeed, a majority position among publishing scholars), it is in the same position as Barthes in *S/Z*.

flexive form, marking a site where the divergences and complicities among ecological constraints, economic infrastructures, and individual desires come strongly into play.

Equivocation

Of the literary theorists discussed in this chapter, Michel Serres is by far the most knowledgeable about the science of chaos. It is central to his project; there is scarcely a chapter of his extensive writing unconcerned with its implications. He uses it, however, in two different ways. In the first, the new paradigms provide him with an implicit vantage point from which to reassess classical concepts. Such a standpoint is necessary, he implies, because we still operate largely within the boundaries of classical thought. Frequently he chooses as sites for this reassessment moments when the classical paradigms were forged, such as the origin of geometry or the formulation of an ideal Platonic object. By showing *what was excluded* at these originary moments, he reveals the complicities that classical constructions share.

In the other mode, Serres attempts to extend the new paradigms to universal theories. Perhaps the best known is parasitism (*The Parasite*, 1982b). Playing on the three senses (in French) of the parasite—noise in a communication channel, an unwelcome guest, and an organism that feeds off its host—he uses its equivocal connotations to create a general theory of exchange. In my view the theory is not successful, for reasons I will indicate shortly. But I find it interesting that such a theory was attempted at all. Why would the theorist who warns us that "the global does not necessarily produce a local equivalent, and the local itself contains a law that does not always and everywhere reproduce the global" (1980: 75) want to create a global theory?

The problem is not peculiar to Serres. As we shall see in chapter 8, it engages many contemporary theorists. I believe that it arises because the more intense the theorist's commitment is to the ideology of the local, the more he feels compelled to strengthen it by making it universally valid. This compulsion leads to a paradoxical attempt to extrapolate a general theory from paradigms that imply there are no general theories. In its paradoxical impulses as well as its subject matter, Serres's project is prototypical of attempts to extend chaos theory to a general theory of culture. Out of its strong internal tensions—one might almost say out of its turbulence—equivocation emerges as a central concept. As different voices compete within the channel of Serres's writing, equivocation serves both as the keystone for his theory of communication and as a metaphor for the conflicting impulses inherent in his interdisciplinary approach.

Serres's commitment to local knowledge is apparent in his essay "Platonic Dialogue" (1982a: 65–70). He points out that writing is inherently global, because it is "first and foremost a drawing, an ideogram, or a conventional graph" (p. 65), that is, a mark recognized as a form independent of its particular manifestation. Experienced readers are so accustomed to screening variations in letter formations from consciousness that they are only subliminally aware of them. But children learning to read frequently notice and comment upon the difference between a printed and a handwritten *a*, or between *t*'s with curved and straight bottoms. To the adult reader, the child's focus on these variations is misguided; yet these "cacographies" (1982a: 66), as Serres calls them, *are all that ever exist.* In learning to suppress local variations, Serres implies, we also blind ourselves to the richly various actualities of the material world.

No less than writing, speech also depends upon suppressing the actual in favor of the ideal. Regional dialects, mispronunciations, stuttering, and slurring are routinely processed not as the sounds they actually are but as the idealized forms they represent. Thus *"the first effort to make communication in a dialogue successful is isomorphic with the effort to render a form independent of its empirical realization"* (p. 69). Through this demonstration, Serres opens a passage between the Platonic dialogues and the development of mathematics. What a Greek geometer meant by a circle, for example, was not the wavering form drawn in the sand but the abstraction toward which the mark was an imperfect gesture. Why, Serres asks, does mathematical discourse even today use the consensual "we" in demonstrations and proofs (p. 68)? Because mathematics is assumed to operate in a realm purged of all accidental, transitory features—that is, a space free of noise. Hence the Q.E.D.: "the dialectical method of the dialogue has its origins in the same regions as mathematical method, which, moreover, is also said to be dialectical" (p. 69). The argument's thrust up to this point is to show how thoroughly the suppression of noise has been incorporated into the foundations of modern thought.

From this demonstration Serres attempts to develop a general theory of communication. A dialogue, he asserts, "is a sort of game played by two interlocutors considered as united against the phenomena of interference and confusion. . . . These interlocutors are in no way opposed. . . they battle together against noise. . . . *To hold a dialogue is to suppose a third man and seek to exclude him*" (pp. 66–67). Although it is not clear why noise should be identified with a "third man," Serres apparently arrives at it by analogy with the excluded middle of binary logic. Just as classical mathematics developed by excluding a third possibility between true and false, so dialogues (and dialectics) evolved by excluding a "third man" who threatens to disrupt communication.

In privileging the "third man," Serres ostensibly proposes to deconstruct the traditional hierarchy of information and noise by inverting it and then revealing the existence of a hitherto unrecognized third term. This deconstructive thrust, however, gives way to the synthetic impulse to make the parasite into a universal concept in its own right. As Serres moves from reinscribing local knowledge to a synthetic theory of noise, conflation is the order of the day. In *The Parasite*, the "third man" becomes a general operator that applies indiscriminately to La Fontaine's fables, thermodynamically irreversible systems, communication channels, and economic exchanges. Such wide-ranging application is possible only because of metaphoric slippage in the half-numerical, half-anthropomorphic "third man." The "third man" can act as an operator in human exchange because it anthropomorphizes noise; it can explain why systems become more complex because it goes by the same name as the formal term that appears in information theory.

With such equivocal definitions, it is scarcely surprising that virtually any exchange can be explained in terms of the parasite. But what does such a theory tell us? Whatever else it intimates, the theory works (if it works) only by assimilating obvious differences into apparent similarities. And yet the explicit intention in creating it was to liberate the noise of empirical differences from the oppression of idealized similarities.

Given this paradox, it is instructive to see how Serres treats the suppression of difference when other people do it. In "Mathematics and Philosophy: What Thales Saw . . ." (1982a: 84–97), Serres takes as his historical site Thales's discovery that he could determine the height of a pyramid by measuring its shadow. He imagines Thales standing in the Eygptian desert before one pyramid; two others loom in the distance, the same and yet somehow different. The mathematical concept of similarity, Serres suggests, is already encoded into the scene. Thales stands "in the domain of implicit knowledge" (p. 89); all he need do is make it explicit. Yet the movement from implicit to explicit knowledge is not inconsequential. It requires that the shadow be negated in its particularity and rendered as the mark of an idealized form. Thus at the birth of geometry, the "shadow of opinion, of empiricism, of objects" is sacrificed to "the sun of knowledge and of sameness" (p. 90).

The components of the scene, Serres notes, are the same as those in Plato's allegory of the cave: sun and shadow, explicit and implicit knowledge, idealized form and empirical fuzziness. No less than the allegory of the cave, the origin of geometry establishes "the scene of representation . . . for Western thought for the next millennium" (p. 92). Only when this "immense historical cycle had finally come to an end" (p. 93) could non-Euclidean geometry and Cantor's "monstrous" sets enter the space of knowledge as legitimate subjects for inquiry. According to Serres, the shift is characterized by the displacement of a linear, chronological aesthetic based on a vision of time passing always at the same pace and measured in the same units by an aesthetic that focuses on spatial deformations and local turbulences.[13]

This line of thought suggests that spatial metaphors will be central to an understanding of the rhetorical moves that allow Serres to celebrate difference in his inscriptive mode while simultaneously

[13]The influence of Mandelbrot is clear here. I suspect, however, that Serres may not really understand the mathematics, for fractal geometry has not renounced globalization by any means (a point discussed in chap. 8).

suppressing it in his general theory. Generally speaking, space for Serres is identified with difference, whereas time connotes sameness. In *Feux et signaux de brume* (1975) he writes, "Time [is] the most immediate and simplest esthetic projection of ordered structure. With time, the esthetic is in order and those in political power are quite pleased. Spaces are repressed because they are possibly, better yet, certainly, disorderly. Reason, the political powers that be, prefer order rather than disorder, time rather than space, history rather than multiplicities" (p. 164). What metaphors does Serres use to liberate space from the "political powers that be"? More precisely, what metaphors allow him to negotiate between the drive toward synthesis in his own writing and the authorized voice that locates this impulse in external political authority?

One is the image of the passage, the journey negotiated with difficulty between narrow straits. In *Hermes V* (1980) Serres appropriates this geographical terrain as a central metaphor for his project, especially for the passages of his writing:

> No, the real is not cut up into regular patterns, it is sporadic. . . . I am looking for the passage among these complicated cuttings. I believe, I see that the state of things consists of islands sown in archipelagoes on the noisy, poorly understood disorder of the sea. . . . Passages exist, I know, I have drawn some of them in certain works using certain operators. . . . But I cannot generalize, obstructions are manifest, and counterexamples abound. [1980: 23–24]

The mediating potential of this image is made clear in Serres's discussion of space in classical Greece (1982a: 39–53). "The Greek cities were dispersed, reciprocally closed insularities . . . in which every man worthy of the name . . . was inside, while on the exterior of this political space animals, barbarians with growling languages, circulated in a chaotic multiplicity of sociopolitical spaces" (p. 51). In this world, when connections are lost or passages not completed, no one "can speak any longer, and we have the irrational or the unspeakable—the incommunicable, to be very precise" (p. 50). The Greeks dedicated themselves to rationality, Serres implies, because universality was a hard-won accomplishment in their dispersed space. When one must fight to establish connections, one does not worry about the tyranny of globalizing theories or universal forms.

In contrast to the "sporadic" space of ancient Greece is the global village Serres inhabits, spanned by supersonic transport and bound by instantaneous satellite communication. Why does he assume this space is "in tatters"? Because the assumption is one of the rhetorical gestures that allow him to reconcile his globalizing impulses with his commitment to local knowledge. "Therefore I assume there are fluctuating tatters," he writes in a statement only apparently at odds with his attempt to build universal theories (1980: 23). In fact, the assertion is what enables the theories to be advanced, for it is only when space is inherently and irrecoverably "tattered" that these theories are distinct from the repression of difference which the powers that be practice. For Serres to assume that his space is dispersed is as audacious as for the Greeks to assume that their space was continuous—and as necessary.

These divided impulses—the desire to proclaim space dispersed and the effort to assimilate difference into similarity—join in the image of the spiral. It is an appropriate image, for in its upward thrust it expresses the yearning for difference; then difference is assimilated into similarity when the trajectory conforms to a circular pattern; yet difference asserts itself again as the spiral continues upward, ad infinitum. Serres's essays are characteristically structured as spirals, coming around again and again to the same point displaced along an axis. For example, in "The Apparition of Hermes: Dom Juan" (1982a: 3–14), the same operators are recycled through slightly displaced locales in Molière's play. "The demonstration begins again" (p. 5), he writes as he considers the first act; "the same demonstration begins again" (p. 8), he repeats a few pages later; once more, the "demonstration begins again" (p. 13), until he finally arrives at patterns so pervasive that he can proclaim them to be "universal" (p. 11).

The classical text that Serres identifies as closest to his world view is Lucretius's *De rerum natura*. It is, he argues, structured like a spiral (1982a: 98–124). Moreover, it envisions the universe as a spiral. "Space and time are thrown here and there. There is no circle. But stochastically, turbulences appear in space and time. And the whole text creates turbulence . . . the creative science of change and of circumstance is substituted for the physics of the fall, and of rigorous trains of events. *Neither a straight line nor a circle: a spiral*" (p. 99; emphasis added). The ideology of the local, never entirely absent in Serres's writing, is explicit in this essay. Had Lucretius's vision of the clinamen prevailed, the world might be dedicated to chaos rather than order, Venus rather than Mars, love rather than death. "The order of reasons is repetitive, and the train of thought that comes from it, infinitely iterative, is but a science of death. . . . Stable, unchanging, redundant, it recopies the same writing in the same atoms-letters. . . . Everything falls to zero, a complete lack of information, the nothingness of knowledge, nonexistence." But when the Lucretian clinamen enters the picture, it "cures the plague, breaks the chain of violence, interrupts the reign of the same, invents the new reason and the new law . . . gives birth to nature as it really is" (p. 100). As we saw in chapter 4, turbulence cannot be adequately modeled through any simple form. Certainly it cannot be modeled as a spiral, which is far too orderly and predictable to express its extreme complexity. Serres does not so much express turbulence, then, as tame it when he envisions the world as a spiral. The strategy suggests that he may be more ambivalent about disorder than his rhetoric indicates.

This ambivalence is writ large in the contrast between the retrospective and globalizing modes in Serres's writing. Returning to sites where global theories were formed, he reinscribes the local knowledge they suppressed; but looking at the world locally, he cannot resist organizing scattered sites into global theories. I conjecture that Serres, despite his professed commitment to the local, keeps driving toward the universal because he cannot help feeling what virtually all scientists and very few poststructuralists do: the power of scientific explanations. His writing does not merely discuss literature and science. Informed both by the expanding economy of poststructualism and by the contracting aesthetic of science, it *is* literature-and-science. Taking equivocation as its central topos, it is also itself deeply equivocal, for the different voices of literature and science are both trying to occupy the same channel at the same time.

In "Lucretius: Science and Religion" (1982a: 98–124), Serres tackles the problem of his divided loyalties head on. Science, he acknowledges, is universal: "Two and two make four; heavy bodies fall, according to the law of gravity; entropy increases in a closed system, regardless of the latitude and whatever the ruling class." How can this universality be reconciled with a proposition he believes just as strongly, that "science is conditioned by postulates or by decisions that are generally social, cultural, or historical in nature?" How, in other words, can science be "conditioned but unconditional? No one has escaped this dilemma," Serres admits, certainly not himself (p. 106). He comes close to resolving the paradox when he argues that the truth content of science is a necessary but not sufficient condition to determine the final form of a scientific paradigm. Reality imposes constraints that cause some hypotheses to be verifiable, others not. But it does not determine which theories are actually realized or how they will be expressed. These things depend on cultural and historical conditions peculiar to the moment. With this lucid insight, Serres has what he needs to open the narrow straits of his passages into easily traversed terrain.

But he does not use it. Instead, the equivocation characteristic of his writing becomes more intense as it comes closer to potential resolution. Consider the sentence that contains the essential insight. It demonstrates in microcosm how intrinsic equivocation is to Serres's rhetoric. Musing on how science can be at once local and global, he writes, "In this case and a thousand like it, *you can always proceed from the product to its conditions, but never from the conditions to the product*" (p. 106). The generalizing impulse is apparent in "always" and "never"; yet these words are preceded by the equivocal phrase "in this case and a thousand like it," which envisions the global as a mass of localized points, not as a true universal. Equivocation is also apparent in the sense of the sentence, which could be paraphrased as "universals apply, but not always." To see whether they apply in a specific case, one has to consider the direction one travels. If one goes backward through time, one can "always" universalize; if forward, "never." Into which category does this statement fall? Since it spans both directions, it must be both true and untrue, local and universal, as its rhetoric signals. In other words, the sentence contains an undecidable proposition. At least since

Russell and Whitehead, we have known how to analyze propositions of this kind. The sentence is undecidable because it is paradoxical, and it is paradoxical because it is self-referential.[14]

To identify the paradox, however, is not to resolve it. The statement is *unavoidably* self-referential, because Serres's entire project is caught between going backward in time when he reinscribes difference in classical paradigms and going forward in time as he strives to create a universal theory of local difference. This divided temporal impulse reflects his divided loyalties toward literature and science. To speak unequivocally would imply choosing one of these polarities over the other; and this he cannot do without vacating the site he has chosen for his project. Thus when he addresses the problem of divided loyalties directly, the prose characteristically becomes more infolded, for only so can the tensions that define the project's site remain intact.

Nowhere are these tensions more apparent than in "The Origin of Language: Biology, Information Theory, and Thermodynamics" (1982a: 71–83). In this essay Serres tries to expand informational equivocation into a general theory of language. After reviewing the second law of thermodynamics, he points out that fighting against entropic decay is the openness of all living systems—their ability to take in sunlight, food, information. Since this flow sustains life, he proposes that we regard ourselves not as stable bodies through which constantly changing streams of matter and energy flow but as stable flows encased within constantly changing bodies. Consider an ocean wave: although the water appears to flow, a typical water molecule moves only a few feet during its entire lifetime. The shores, by contrast, are eroding every day. In an elegant reformulation of Heraclitus, Serres writes, "One always swims in the same river, one never sits down on the same bank" (p. 74).[15] This part of the essay is

[14]Serres's views on this matter are very similar to those of Lem, who also positions himself at an interstice between literature and science. It is no accident that Lem, like Serres, sometimes opts for a transcendent viewpoint rather than a self-reflexive loop. His topological model, for example, must of necessity be one of the shapes possible within postmodern culture. But Lem does not inquire into the conditions of possibility that authorize his topological model of culture, preferring to posit it as a transcendent concept rather than as a model that models itself.

[15]Serres may not be quite correct here. It is true that waves in an ocean serve to transfer energy rather than matter, but I rather think that water molecules in rivers

in the retrospective mode, an attempt to insert into traditional paradigms the empirical differences they suppress.

Having allowed the voice of difference to speak, Serres then moves to assimilate it into a general theory of language. His inspiration is the article by Henri Atlan (1974) cited in chapter 2, in which Atlan shows that the value of the equivocation in Shannon's equations depends upon where the observer is positioned in relation to the message. If the observer knows what the "correct" message is—that is, if she is positioned at the message source—the equivocation will appear as unwanted noise that should be subtracted from the message's original information. If the observer is positioned outside the system, however, she is in a position to notice the effect of the equivocation on the system as a whole. She thus can see instances in which the equivocation causes the system to reorganize itself at a higher level of complexity. In these instances, Atlan argues, the equivocation should be given a positive sign, for it has in effect become information. It is not hard to see why this argument would appeal to Serres (to say nothing of Barthes, who would find it deliciously redundant). Serres correctly sees in Atlan's article scientific validation for his view of noise as a positive force.

However, he continues to a conclusion found in none of his scientific sources. He apparently arrived at it by conflating Atlan's article with François Jacob's speculation (1976) that biological organisms are structured in interlocking levels of integration, like so many Russian dolls stacked inside one another. If organisms are structured in hierarchical levels, Serres reasons, then one can conclude that the value of the equivocation changes with each level, so that each higher level functions as a "rectifier of noise" for the level beneath (1982a: 78). If this were so, noise at a lower level would always be transformed into information at the next higher level, and organisms would be unified only *because* they are fragmented—an implication Serres expands into the paradoxical conclusion that language itself originates in noise.

At this point, his divergence from his scientific sources is clear.

---

do move. Otherwise, as a colleague pointed out, why would mountain streams run dry?

There is no scientific evidence that noise is rectified from level to level within human beings. Nor do these theories say that noise *necessarily* becomes useful information. Whether noise will have a positive or negative effect on systemic organization depends on the stability of the system, the kind of feedback loops at work, the amount and kind of noise injected, and when the injection occurs (Atlan, 1974; Jacob, 1973; Nicolis and Prigogine, 1977). Serres's claim that the rectification of noise "is valid for all levels" and is a universal "law of the series" is simply not true.

For Serres, however, this conclusion is merely the springboard for further speculation. He links equivocation with the Freudian unconscious, reasoning that if noise can be either positive or negative, then one can say that "in one case, it covers up; in the other, it expresses." Thus he concludes that the "entire symbolic function is embedded in this process, the entire strategy of free assocication, Freudian slips, jokes and puns" (p. 80). As he diverges further from fact, his rhetoric becomes more insistent that his theory is scientific. "These matters are straightforward" (p. 78), he claims, alleging that "we can speak about [a layered unconscious] by using a discourse which ultimately can be expressed in mathematical terms" (p. 79). To whom does the "we" in this sentence refer? Not to the sources Serres cites—Henri Atlan, François Jacob, and Claude Shannon, for example—or to any other scientist of whom I am aware.

If we think about what is happening at this point in Serres's argument, we can see that he is trying to make the rectification of noise into a principle that applies equally to language, human psychology, the universe, everything. The irony is that in transforming equivocation in this way Serres has made it univocal, for it leads always to increasing order. Thus "rectified," equivocation is at least as capable of suppressing difference as any of the Platonic concepts Serres addresses. Projecting this no-longer-equivocal equivocation onto an imagined scientific "we," he dreams that it can fuse the deep shadows of the unconscious with the clear sun of mathematical equations, just as he dreams that his own discourse is at one with all other discourses, from Plato to Freud to Shannon. Of course these dreams are not realities. But they nevertheless represent a culminating vision, as the heightened language signals, for at last the noise of difference has been made to speak the language of unity. Serres imagines that through his theory we will finally understand the "packages of chance" that come "crashing at our feet, like the surf at the edge of the beach, in the forms of eros and death" (p. 80). It is from this aporia that his theory of the parasite emerges, traversing a path that can be negotiated only in the equivocal language of dreams.[16]

In one sense Serres is a man ahead of his time, for he understands that the reevaluation of chaos within contemporary paradigms is a cultural shift of the first importance. It signals not just a new scientific or literary theory but a shift in the ground of representation itself. However, the passage between wacky theorizing and brilliant insight in his writing is so narrow that it is sometimes hard to say on which side it falls. It seems most uncontrolled when it takes itself as subject, that is, when it operates upon the equivocation that is its abiding subject and underlying dynamic.

Then the strain of self-reflexivity, which I noted earlier as an undertone, is exaggerated until the prose is caught in the recursive loops typical of positive feedback, oscillating more and more wildly between the polarities that define its dialectic, insisting all the while that this oscillation is really unity. Self-referential paradoxes are, of course, the stuff of which deconstructive analyses and postmodern literature are made. Within the context of Serres's writing, however, they have the effect of conflating the distinct voices of literature and science into a cacography of confused claims. This transformation of equivocation into cacography signals failure, for Serres's project depends upon speaking both the claims of science to represent the world and the admission of language that it can represent only itself.

[16]At the end of *The Parasite*, Serres makes the significant gesture of rejecting the ocean-born Venus for the ocean itself. The gesture marks a new phase in Serres's development, in which he increasingly abandons all forms of order to crawl down the "manholes" or "bubbles" in the foam to commune with aboriginal chaos. These aspects of Serres's more recent work are discussed by Eric White in "Negentropy, Noise, and Emancipatory Thought" and by Maria Assad in "Michel Serres: In Search of a Tropography," both in *Chaos and Order: Complex Dynamics in Literature and Science*, ed. N. Katherine Hayles (Princeton: Princeton University Press, forthcoming).

In attempting to articulate literature and science together through the three modes of interaction discussed in this chapter—ecology, economy, and equivocation—I have sought to create an equivocal site at which both disciplines can have a voice. The vision I hope to have conveyed is not of science influencing literature but of literature and science as two mingled voices within the cacophonography that we call postmodern culture.

CHAPTER 8

The Politics of Chaos: Local Knowledge versus Global Theory

In the late twentieth century, local knowledge has come into its own. Across a wide spectrum of contemporary theory—in philosophy, feminism, literary criticism, and cultural analysis—theorists insist that local variations must be respected in their own right and not simply incorporated into global schemes. The issue is made more complex by the shifting meanings that "local" and "global" carry in these arguments. In critical theory, for example, "global" is understood to mean not only cultural systems considered as a whole but any theory that subsumes particular texts or phenomena into a universal explanation. Marxism, relativity theory, and grammar are global theories in this sense. Similarly, "local" connotes both a small subsection of a geopolitical area, such as the western coast of Brittany, and particular textual or cultural sites that resist assimilation into the generalizations of a universal theory.

The conflation of geopolitical with theoretical connotations is significant, for it signals a growing feeling that totalizing theories should be discredited because they are associated with oppressive political structures. Particularly important here are Foucault's archaeological analyses of the totalizing theories of the Enlightenment, from grammar to biology to penology, and their complicity with totalitarian political practices (1970, 1973, 1977). By implication, local knowledge has become associated with liberation. I shall look critically at these assumptions, arguing that local knowledge does not *necessarily* lead to liberation, any more than generalizing theories always lead to oppression.

The new scientific paradigms, especially the science of chaos, have been seen as carrying the case for local knowledge into the physical sciences. This argument is also suspect, for reasons I will discuss shortly. It is true, however, that the sciences of chaos make the local/global relation problematic in a way it was not in older paradigms. With the onset of chaos, different levels tend to act in different ways, so that locality intrudes itself as a necessary descriptive feature, defeating totalization. To illustrate this process for fractals, Mandelbrot asks a question that looks as if it should have a straightforward, factual answer: How long is the coastline of Britain? (1983: 25–33). As I intimated earlier, the question is more devious than it appears, because the answer is scale-dependent. If we use a mile-long ruler to measure the coastline, we get a shorter answer than we do if we use a yardstick, for the mile ruler cuts across irregularities that the yardstick measures around. If we use an inch ruler the answer is still longer, because small pebbles are measured around; and if a micrometer is used, even irregularities within a single pebble count. In fact, Britain's coastline *continues to grow without limit* as the ruler scale decreases, at least down to molecular scales. Unless the length of the ruler is specified, the question cannot be accurately answered.

The example illustrates how questions of scale are foregrounded in the new paradigms, in contrast to Euclidean geometry and Newtonian mechanics. In these older paradigms the idea that one could get different answers with different scales does not appear. It is not quite correct to say they make global statements that are considered true at every level, because their globalizing approaches are so complete that the system is not conceived as having levels in any meaningful sense. In Euclidean geometry, for example, it does not matter whether an isosceles triangle is twice as large or two hundred times as large as another triangle of the same shape; whatever the scale, Euclidean geometry states, the three sides of similar triangles will be in the same proportion to one another. Fractal geometry does not challenge this assertion as such. Rather, it shifts the focus to complex irregular forms, for which scale appears as an important consideration and movement between scales is highly nontrivial. Similarly, Newtonian mechanics applies the same partial differential equations whether the object is a golf ball or a planet; in either case the masses move through time according to uniform laws of motion. However, when the object has a complex internal structure consisting of distinct local levels, such as a fluid in turbulent flow, the length of the scale is critical because different portions of the fluid move at different speeds and with different kinds of motions. Under these conditions, Newtonian-based calculations are unmanageable for even a few points, and unthinkable for the thousands or millions it would take to model the system.

It is important to understand that chaos theory does not renounce globalization. Rather, chaos theory achieves globalization in a different way, by correlating movements from one level to another. We saw in chapter 7 that a standard method for analyzing complex systems is to look for shapes or configurations having the property of self-similarity. Self-similarity is an essential characteristic because it is *only* when complex systems have this property that scales of different lengths can easily be correlated. The distinction between the classical and new paradigms is not, then, that one globalizes and the other does not, but that one is scale-invariant and the other is not. In this respect chaos theory extends the lessons of quantum physics and relativity theory. After Einstein, Bohr, Planck, and Schrödinger, physicists learned to specify that Newtonian equations applied only at nonrelativistic speeds and to distances and masses where quantum effects could be ignored. Quantum mechanics and the special theory of relativity thus introduced scale considerations—but only for the very small and the very fast. Chaos theory, by contrast, teaches that scale is *generally* important for complex systems, even at nonrelativistic speeds and for macroscopic dimensions.

Chaos theory has also changed expectations about proportionality between cause and effect. In classical paradigms, a small cause is generally associated with a small effect. If the angle between the sides of a triangle changes by one degree, for example, the resulting form is only a little different from its predecessor. In a complex system, by comparison, small changes in the iterative formulae or initial conditions can result in very large changes in the final state. As we saw earlier, this complex relation between the system's behavior and its analytical description is a function of its infinitely fine-grained structure, with symmetries on one level giving way to symmetries on the next, and so on down. The large number of coupling points between levels means that tiny initial changes can quickly be magnified and brought up to macroscopic expression. The sense that movement between the local and global is highly nontrivial thus applies both to levels within a system and to the correlation between the system's physical properties and its analytical description.

In the problematic relation chaos theory posits between description and behavior, it comes close to the critique of global theories within the human sciences. Consider again the question of Britain's coastline. In addition to demonstrating that the answers to such questions are scale-dependent, Mandelbrot shows that they have a political dimension. Portuguese encyclopedias, for example, report that the boundary Portugal shares with Spain is longer than Spanish encyclopedias say it is, by as much as 20 percent. Mandelbrot conjectures that as a small country, Portugal is more concerned than Spain is with the length of its borders and consequently uses a smaller scale to measure them. The example goes to the heart of the objection to global theories within critical theory: to what extent are global theories social and linguistic constructions, inventing the reality they purport to describe? If the answer is "entirely, or nearly so," then the political and ideological functions global theories serve should be the focus of our inquiry, rather than the validity of the theories themselves.

The critique of global theories, then, proceeds on two counts. In the first instance, global theory is questioned because it is not applicable to complex systems. Clifford Geertz, for example, insists on the superiority of local knowledge over global theory because he regards cultures as complex semiotic systems organized around particular local sites (1973: 3–32). In Geertz's view, when one talks about culture in general, one misses the very essence of what one hopes to understand, for meaning cannot be separated from the particular organization of signs that characterizes a given site. In the second instance, globalizing theories are rejected because they are constructions that serve the vested interests of particular classes or power structures. Local knowledge is superior to global theory because, far from reinscribing differences that have marked oppressed people as inferior, it reveals fissure lines marking the interests of the oppressed as different from those of the people in power.

No one would debate the importance of this work. To see the oppression that global theories have too often reinforced or enabled, it is enough to remember how a few European cultures have been equated with mankind; how mankind has been equated with humankind; how humankind has been equated with intelligent consciousness. As only one example of the powerful analyses to which the privileging of local knowledge gives rise, consider Mary Poovey's article (1986) on the nineteenth-century debate within the medical community over whether ether or chloroform should be the anaesthetic of choice. Poovey demonstrates that this seemingly objective question in fact reveals fissure lines that follow the division of power between male midwives and surgeons, both fighting to establish their legitimacy against female midwives and to capture the lucrative practice associated with medical intervention in the birth process. The anaesthetized body of the woman is the blank space that each group textualizes according to its interests. Clearly, these totalizations do not reveal the experiences actual women had, much less the universal truth about such experiences; rather, they are the constructions that the two sides found expedient for their purposes. If either the medical discourse or its image of woman is seen as monolithic, the rifts that mark the play of power are hidden from view. In this instance as in countless others, attending to difference opens discourse to the kind of ideological analysis that can unmask totalizations for the political constructions they are.

Given the power and importance of this method, it is easy to see how the case for difference can be viewed as irrefutable. Under these circumstances, who would want to criticize local knowledge or endorse global theory? If we regard the case as closed, however, we paradoxically blind ourselves to differences within the wide-ranging debate about the local and global in contemporary thought. In the arguments that follow, I try to open this play of difference by drawing distinctions between what "local" and "global" mean in the sciences of chaos and in critical theory. In the new scientific paradigms, the global subsumes the local, but at the price of reconceptualizing the global as constituted by locality. Within critical theory, the claims of the local are expanded until the local itself becomes a new kind of globalizing imperative.

These two impulses mirror each other, for in the sciences of chaos the global is localized, and in critical theory the local is globalized. If the local and global turn themselves inside out when pushed to extremes, how can their conflicting claims be adjudicated? In searching for ways to articulate them together, I argue that it is wrong to assume that global theory is always politically more coercive than local knowledge. For example, the abolitionist argument that slaves had souls was a global theory about human nature; yet it could hardly be said to be more coercive than local knowledge about slaves in the South.[1] The example demonstrates that the case for local knowledge, to be meaningful, has to be historically specific. To argue that local knowledge is *always* better than global theory is, in effect, to create a global theory of local knowledge. Conversely, to ignore the claims of the local is to regress to totalizing theories that have historically been associated with repression. How to find a balance? The question is inherently paradoxical, for to answer it one must put forward generalizations, yet generalizations are precisely what are at issue. The question is also highly charged, for of all the issues energized by the new science, it is the most inescapably political.

Why should there be a balance? This prior question goes to the heart of the issue, for the local/global dialectic is deeply tied up with the long-standing debate between the physical and human sciences about the extent to which reality is a social construction. If there is only epistemology and no ontology, universals within physical theories reflect only how we perceive, never reality itself. In this view, there is no difference between the second law of thermodynamics and the doctrine of white supremacy; both are social and political constructions that serve the interests of specific groups of people at specific times. In my view, this statement is both true and false. It is true in the sense that all knowledge, including the second law of thermodynamics, is representational. It is false in the sense that some representations of the physical world can express universals that are

[1]My thanks to Donald N. McCloskey for suggesting this example to me.

true at many locales. As Michel Serres puts it in *Hermes* (1982a), "Entropy always increases, whatever the latitude and whoever the ruling class." Finding a balance implies finding a way to say that entropy always does increase, at least in this universe, and also that entropy is a social and historical construction that has meant different things to different people for different reasons. If I seem more critical of local knowledge than of global theory, it is because in the last decade the ideology of local knowledge has been too often assumed and too little examined. My purpose is not, however, to criticize local knowledge as such, but to explore its limits so as to allow also for the possibility of global theory.

The ideological freight carried by the valorization of local knowledge is explicit in Jean-François Lyotard's conclusion to *The Postmodern Condition* (1984). He foresees that the rapid growth of teletronics in highly developed societies will further consolidate power in the hands of the elite who have access to data banks. He implies that this trend can be countered by the emergence within science of fractal geometry, quantum mechanics, catastrophe theory, and Gödel's theorem. Grouping these disparate theories under the label "paralogy," Lyotard suggests that they will let us "wage a war on totality; let us be witnesses to the unpresentable; let us activate the differences and save the honor of the name" (p. 82). Here, then, the new paradigms are enlisted under the banner of local knowledge to neutralize the totalizing potential of modern information technology. The argument that "paralogy" can rescue postmodern culture from totalitarianism is ill founded for several reasons. It is akin to social Darwinism, in that it confuses scientific theories with social programs. But even if this were not so, the argument still ignores the fact that the global is not absent from these theories, only redefined by them. The endorsement of local knowledge within contemporary science cannot be the panacea Lyotard imagines. It is true that the new paradigms recognize the importance of scale, and therefore of locality. But it is also true that these changes are located within disciplines committed to universal theory. These disciplinary commitments to universality inform the theories as surely as the theories help to form the disciplines.

The universalizing impulse within chaos theory is hard to miss. Surely chaos theory would not have attracted the attention it has if it simply confirmed the obvious, that chaos is disordered. No: what makes it noteworthy is the discovery of order in the midst of disorder. In Feigenbaum's study of iterative functions (1980), the result that makes his work ground-breaking is his discovery that chaotic systems, despite their inherent unpredictability, share certain universal characteristics. The name he chooses to describe his discovery—"universality theory"—clearly conveys this message. The article on strange attractors by Crutchfield, Farmer, Packard, and Shaw in *Scientific American* (1986) is another case in point. For Crutchfield and his collaborators, "chaos" connotes a *patterned* unpredictable trajectory in phase space. If a system is not a strange attractor—if it is simply disordered and nothing more—they consider it a "featureless blob" of no interest (p. 56). Even Mandelbrot (1983), though his rhetoric celebrates disorder, looks for the symmetry groups that will allow chaos to be understood as iterative extensions of regular forms. Kenneth Wilson's Nobel Prize work (1983) on renormalization groups in quantum field theory has the same goal, to find the symmetry operators that permit analytical solutions to chaotic systems. As these scientists use "chaos," it connotes not the unpredictable aspects of disordered systems but their universal characteristics. To see chaos theory as the antidote to totalization is to ignore this thrust toward universality.

Other instances of Lyotard's "paralogy," such as quantum mechanics, are similarly misrepresented when they are enlisted in the cause of local knowledge. Since about 1960 the most active work in quantum mechanics has been the attempt to combine it with relativity theory to create a grand unified field theory. Far from endorsing local knowledge, grand unified theory aims to explain all material interactions in terms of the four basic forces of the universe (Parker, 1986; Feynman, 1986). How deeply the universalizing impulse is felt within contemporary science is also apparent in the recent work on supersymmetry, which promises to synthesize the four dimensions of spacetime with electricity and magnetism to create models of the universe ranging from ten to twenty-six dimensions (depending on what assumptions are made) capable of explaining all known physical fields through a single theory (Davies, 1984). To instance quantum mechanics as an example of local knowledge, then, is to skew the way it is used in its scientific context.

At the same time, the emphasis on locality within the new paradigms has changed the way the global is conceived. The essential concept in this changed view is self-similarity. Many complex forms have smaller parts that are shaped like the object—a twig reproduces the shape of a branch, and a branch reproduces the shape of a tree. In these instances the whole and the parts are related through self-similarity. Self-similarity has made it possible to encode the information contained in complex forms in new ways. Michael Barnsley's recent work on fractals, for example, starts with an approximation of the desired global shape—say, the rough outline of a fern (Barnsley et al., 1986; Peterson, 1987). Each frond coming out from the fern's central stem is shaped like the fern. These fronds can be sketched through a series of reducing and deforming transformations (known as affine transformations) that operate upon the fern's global shape to create the individual fronds. The computer re-creation can be as detailed as desired. Individual leaflets on a frond, which are shaped like the fern, can also be generated through affine transformations; so can smaller irregularities on the leaflets, and yet smaller variations in these irregularities. In this way, increasingly detailed structures are obtained from the same initial bits of information. The method has considerable generality and can be used to recreate such diverse natural forms as landscapes, capillaries, clouds, and galaxies. It allows complex images, which would normally require several million bits of information to describe, to be encoded through the few iterative formulae used to generate the self-similar shapes and carry them through the transformations. Some researchers predict that in a few years, movies could be transmitted from computer to computer by this method (Peterson, 1987: 283).

As Mitchell Feigenbaum and Benoit Mandelbrot have pointed out, such methods have important epistemological implications.[2] In the Newtonian paradigm, objects moving through time are modeled as trajectories that can be broken into arbitrarily small intervals (this is essentially the basis for calculus). As these intervals approach zero,

[2]I am indebted throughout this discussion to the presentations by Mitchell Feigenbaum, "Attempts at the Analytic Description of Complex Phenomena," and Benoit Mandelbrot, "Fractals: Complex Relationships in Mathematics, the Sciences and the Arts," at the conference "Analyzing the Inchoate," Cornell University, April 16, 1987.

they are added together to get the trajectory as a whole. Underlying this method is the common-sense perception that movement through time can be equated with the movement of an object along a predictable path, such as the parabolic arc of a baseball as it comes off the bat. However, if the moving shape is complex in the sense of being composed of multiple levels acting in different ways, the calculations become unmanageable very quickly, because each of the levels has to be advanced through time individually. In these cases, it is much more economical to describe the transformation rules that govern the evolution of the iterative formulae through time. The shape itself, moreover, is no longer conceived as a mass of points, but as the formulae used to generate and randomize the self-similar forms that compose the various levels of the object. In effect, the underlying epistemology has changed radically, for the system no longer is conceived as masses of points moving along a predictable path, but is seen as the evolution of the internal structural principles that describe the propagation of self-similar symmetries.

Consider how this change in epistemology would affect traditional paradigms within the human sciences. Newton's conception of objects as masses of points is analogous to Hobbes's vision of society as a group of autonomous individuals, and to Adam Smith's representation of the economy as a collection of competing customers. In these conceptions, the individual units that collectively constitute the global system are considered to be elemental points acting in accord with general laws. One makes the transition from the local to the global by applying general laws to masses of individual units, and one achieves movement through time by adding together the individual motions to arrive at a resultant. Now compare these classical assumptions to those Foucault uses in his archaeologies. Foucault considers the individual not as an autonomous point but rather as a microcosm constituted by the tropes and organizing figures characteristic of the episteme. For Foucault, individuals do not constitute culture; culture constitutes individuals. Moreover, the concept of the episteme implies that different sites within a given cultural period are self-similar. Foucault (1970) finds that during the age of classical reason, for example, the same organizing tropes appear in grammar, biology, political theory, and psychology.

A criticism frequently made of this approach is that it does not explain how changes take place through time (assume, for the purposes of this argument, that Foucault has characterized the periods accurately). How and why did people stop believing in similitude as an organizing trope and start believing in representation? The objection is not so straightforward as it appears, for what counts as a temporal explanation in the new paradigms is fundamentally different from what counts in the old. In traditional models, a temporal explanation placed groups of autonomous actors along a time line and advanced them according to laws or generalizations that explained why the actors behaved as they did. In the new paradigms, by contrast, temporal explanation connotes an understanding of the structural principles that relate different sites together by self-similarity, along with rules that state how these principles evolve over time. Foucault has achieved the first step in this kind of explanation by anatomizing the structural principles that underlie self-similar forms. Although he has not given a full temporal explanation, he has nevertheless come closer to it than his critics acknowledge, because they discount comparison of different kinds of self-similarities as the foundation on which temporal explanations are built.

Foucault's work suggests that the new paradigms could provide a model for culture, as well as for disciplines within the culture. Recall that in the sciences of chaos, the local designates the site within the global at which the self-similarities characteristic of the system are reproduced. Conceived as images of each other, local and global are related as microcosm is to macrocosm, although each level also contains areas so complex that they are effectively chaotic. Movement between levels is easy or possible only when the symmetries align. Configurations that replicate symmetries are ripe for change, because at this point the macroscopic properties of the system are extremely sensitive to microscopic perturbation. Applied to cultures, such a model does not require that all configurations within a given culture be self-similar; but it predicts that when enough of them are, initial perturbations will have large-scale effects. From the viewpoint of the cultural analyst, the model offers several advantages. Because they occur at multiple sites and many levels, characteristic symmetries are relatively easy to document; and because the system at this point is susceptible to dramatic and sudden changes, the changes that do occur are easy to recognize. The model gives us a way to understand how Foucault could be right in postulating a self-similar episteme for a given period, and still does not commit us to saying that knowledge everywhere in this period was organized according to these tropes (because there are always random variations) or that culture always works like this. In other words, it gives us a way to say that the local and global sometimes reproduce each other so that universals appear to run through the system, and sometimes not.

The logical next step is to wonder about the episteme of which this model is characteristic. There is a self-reflexive turn to this thought which makes it hard to think, for the model is used to explain how different cultural sites make up an episteme, and the episteme is used to explain how such models could arise at multiple sites within the culture. In other words, the mark of our episteme is that the concept of an episteme is a thinkable thought. The circularity of the proposition may explain why Foucault, especially in his earlier texts, emphasizes that one can never know the episteme in which one lives. By making this claim, he forecloses the self-referential circling that would otherwise engulf the episteme of epistemes.

Is it true that one's own episteme must remain on the dark side of knowledge? On occasion Foucault seems to imply that since the episteme is the ground of knowing, it cannot itself be known, at least not by those who stand upon it. However, this argument underestimates the heterogeneity of culture.[3] Older epistemes do not disappear; rather, they continue as substrata valid for organizing restricted spheres of experience. When I drive to the store or play volleyball, for example, I am a Newtonian, never doubting that the laws of motion apply unequivocally. I know this not only intellectually but kinesthetically and intuitively. When I wonder why the price of gasoline has risen ten cents this week, I contemplate the complex mutual interactions most economically expressed through the field concept that also serves as a metaphor for global society. When I watch fractal forms being constructed on a computer screen, I am aware of yet another kind of thinking, distinct from the other two, with a wide domain of applicability to natural forms and complex systems. In what episteme do I live? Not in any single epis-

[3]In Foucault's later work, especially in vol. 1 of *A History of Sexuality* (1980a), it becomes clear that he sees earlier epistemes as forming the substrata of later periods.

temology, but in a complex space characterized by multiple strata and marked by innumerable fissures.

Our episteme is visible to us, at least in part, because it is not homogeneous. Stratified through time and fissured into different sites, its formations are various enough to provide multiple perspectives. By standing at one perspective—for example, at the prospect offered by the sciences of chaos—one can see more clearly other perspectives, such as that of critical theory. Yet the metaphor also implies that different sites are underlaid by a common ground. The similarities between critical theory and the science of chaos justify the claim that they constitute a common epistemology, if not an episteme. At the same time, there are rifts that reveal significantly different assumptions at work, which is another way of saying that these disciplines are located at different sites.

The interplay between difference and similarity, critical speculation and scientific theory, brings us back to the central issue. Can any totalizing theory be true, when we know that our conception of the world is dependent on the local particulars of our lives? Are the claims of science to represent an objective reality justified, in the face of all we know about how linguistic, cultural, and physical factors affect our vision of the world? We saw in chapter 7 that Serres attempted to answer these questions by recognizing the difference between prediction and retrodiction. I will pursue another tack, arguing that we can adjudicate the issue by recognizing that reality cannot be identified in its positivity but must be inferred from the constraints applicable in a given situation.

Consider, for example, how conceptions of gravity have changed over the last three hundred years. Gravity is conceived very differently in the Newtonian paradigm in the general theory of relativity. For Newton, gravity was the result of mutual attraction between masses; for Einstein, it was the result of the curvature of space. One might imagine still other kinds of explanation, such as a Native American belief that objects fall to the earth because the spirit of Mother Earth calls out to kindred spirits in other bodies. But no matter how gravity is conceived, no viable paradigm could predict that when someone steps off a cliff, she will remain spontaneously suspended in mid-air. This possibility is ruled out by the nature of physical reality. Although the constraints that lead to this result are interpreted differently in different paradigms, they operate universally to eliminate certain configurations from the range of possible answers. Gravity, like any totalizing idea, is always and inevitably a representation; our entire perceived world consists always and only of representations. Yet within the representations we construct, some are ruled out by physical constraints, others are not.[4]

This position should not be misconstrued as positivism. Positivism was part of a foundationalist movement that sought to establish an unambiguous connection between theory and observation, knowledge and perception. To remove ambiguity and to demonstrate the *necessary* nature of the connection between theory and observation, positivism turned to formalization. The purpose of formalization was to ensure that no subjective elements entered into the transformation of observation to theory. If one assumed that the scientific method reliably screened valid observations from suspect data, one then had a carefully forged chain that stretched from the foundation of facts to the superstructure of theory.

Today positivism is often evoked in literary studies as a synonym for naive realism. Its position with respect to realism is ambiguous, however, rather like the position of relativity theory with respect to classical physics. In one sense, relativity theory is the culmination of classical physics, because it is strictly causal and dedicated to preserving the universality of physical laws. In another sense, it marks a rupture with classical physics, for it demonstrates that the Newtonian paradigm is applicable over only a limited range of conditions. Moreover, it transforms the meaning of space and time even within the limited range that Newtonian mechanics covers. Similarly, in one sense positivism is an attempt to extend realism and make it more rigorous. In another sense, the very need to inject rigor indicates that realistic premises, by themselves, are shaky. Only when the chain from observation to theory is perceived to be weak is it necessary to strengthen it.

There is no need to rehearse here the well-known story of how and why positivism failed. Quantum mechanics, Gödel's theorem, and the sociology of science collaborated to show that causality in

[4]This argument obviously has certain affinities with Karl Popper's arguments regarding falsification (1965:33–65), although Popper is committed to a stronger version of realism than my position requires.

the physical world is problematic, that formalization can never be complete, and that observations are always shaped by preexisting assumptions. The failure of foundationalist approaches across a range of disciplines led to an era of reflexivity, when artists, philosophers, writers, and scientists became fascinated by the paradoxical possibilities of a mind that watches itself thinking that it is thinking, of a left hand that draws a right hand drawing the left hand, of a text that writes a text about writing a text. This reflexive turn of mind is the parent, or at least the stepparent, of the vacuum that produces something out of nothing. The fecund vacuum would not be the powerful image it is in contemporary thought if reflexivity had not prepared the way by intimating that a lack of ground can be productive and exhilarating rather than threatening. Even if there are no foundations, there may nevertheless be creation, evolution, and renewal.

Consider how the position I articulated earlier compares with positivism. The proposition that we are always already within the theater of representation assumes at the outset that no unambiguous or necessary connection can be forged between reality and our representations. Whatever reality is, it remains unknowable by the finite subject. The representations we construct are determined by such historically specific factors as prevailing disciplinary paradigms and cultural assumptions, as well as by such fundamental givens as the human sensory apparatus and neurophysiology. Observations, then, are culturally conditioned and anthropomorphically determined. Thus we can never know how our representations mesh with reality, for we can never achieve a standpoint outside them.

Nevertheless, within the range of representations available at a given time, we can ask, "Is this representation *consistent* with the aspects of reality under interrogation?" If the answer is affirmative, we still know only our representations, not reality itself. But if it is negative, we know that the representation does not mesh with reality in a way that is meaningful to us in that context. The difference between a representation that is consistent with reality and one that depicts reality is the difference between a metaphor and a description. In this sense all theories are metaphoric, just as all language is.

The position I am describing abandons all attempts to arrive at foundationalist certainty. On the contrary, it grants that the connection between reality and representation is essentially unknowable. It emphasizes that reality cannot be known in its positivity, only partially sensed through the failure of some representations to mesh with some aspects of reality. Because its thrust is toward explanations that are consistent with reality rather than reality itself, it could aptly be called tropism. Unlike positivism, it invites cultural readings of science, for the representations presented for disconfirmation have everything to do with prevailing cultural and disciplinary assumptions.

An omniscient viewpoint is clearly not appropriate to tropism, for the implication that reality can be seen completely and transparently is fundamentally at odds with the obscurity tropism posits between reality and representation. A better metaphor than vision is touch, for in contrast to sight's universalizing tendencies, touch is always localized.[5] Not only is it limited to specific body surfaces; it is also limited in the portions of an object it can apprehend. As the fable about the three blind men touching an elephant suggests, the difference between seeing and touching is the difference between a totalizing concept and competing local representations. We can say, then, that in our brushes with (not view or vision of) reality, it cannot be seen, but it can sometimes be felt as our representations push against the limits of the possible. From this position, we can grant the enabling premise of science, that reality can be felt; we can also grant the enabling premise of cultural analysis, that what we see is always and only a representation, never reality itself.

If such a common ground exists, why does contemporary critical theory so often see (or feel) itself opposed to science? Undoubtedly there are many reasons. One is the continuing imbalance between the power and prestige of science and those of other disciplines. As long as science is accepted as the arbiter of reality, other disciplines will resist by empowering their own inquiries at its expense. Also

[5]In "Towards a Critical Regionalism," in Foster, 1983, Kenneth Frampton makes a similar proposal. He writes, "The liberative importance of the tactile resides in the fact that it can only be decoded in terms of *experience* itself: it cannot be reduced to mere information, to representation or to the simple evocation of a simulacrum substituting for absent presences" (p. 28). The claim that touch cannot be encoded as information is probably not correct; for example, Mandelbrot has proposed fractal dimensionality as a measure of surface roughness. But this is a quibble. The thrust of the comment, that touch has historically not been associated with abstraction, is true enough.

important is the lingering of logical positivism within the scientific community. In the eyes of the logical positivists, what I have been calling the common ground is the muck of ambiguity; and as long as this muck is seen as something to be cleaned up through formalization rather than as a problematic to be explored, cultural analysts will see the scientific method as inimical to their projects. Still other factors are the differences in the criteria for publishable work in the two fields. It is possible to be a productive scientist and never seriously consider the problematics of representation. The situation is very different in contemporary critical theory, where questioning the nature of representation is virtually a necessity for work in the field.

As important as any of these factors, however, is the different roles that constraints play in the two fields. For literature, and to some extent for the human sciences, the important constraints are social rather than physical.[6] For example, it would be possible to demonstrate that *Hamlet* is a dramatic text rather than a place mat or a name tag by refererence to the physical constraints that distinguish these forms. But of course no one is interested in making this kind of argument. For arguments people are interested in making—that our interpretation of *Hamlet* is constrained by a community of readers, for example—the constraints are social rather than physical and hence are themselves social constructions. How does one define a community of readers? How does one know what attitudes they hold? Because of the social nature of literary constraints, it is tempting for literary theorists to believe that physical constraints are also constructions, and thus incapable of unambiguously indicating whether a physical theory is false.[7]

[6]For this insight I am indebted to conversations with Mark Seltzer.

[7]Lately this position has also been argued by some philosophers of science, although it has not made much headway among scientists themselves. For a persuasive argument that experimental replication is a social construction, see H. M. Collins, *Changing Order: Replication and Induction in Scientific Practice* (Beverly Hills, Calif.: Sage, 1985). Also relevant is B. Barnes and D. Edge, eds., *Science in Context: Readings in the Sociology of Science* (Cambridge: MIT Press, 1982). Gabriel Stolzenberg carries the argument into mathematics by pointing out that the sense in which a proposition has been "proved" is dependent on prior and unexamined language usage, in "Can an Inquiry into the Foundations of Mathematics Tell Us Anything Interesting about Mind?" in *Psychology and Biology of Language and Thought: Essays in Honor of Eric Lenneberg*, ed. George A. Miller and Elizabeth Lenneberg, pp. 221–269 (New York: Academic Press, 1978).

In my view, the difference between physical and social constraints lies in the fact that physical constraints manifest themselves in isomorphic ways in different representations, whereas social constraints are specific to the representation within which they occur. The present limit on silicon technology, for example, is a function of the speed with which electrons can travel through the semiconductor. One could argue that "electron" is a social construction, as are "semiconductor" and "silicon." There is an unavoidable limit inherent in this constraint, however, and it will manifest itself in whatever representation is used, provided it is relevant to the representational construct. Suppose that the first atomic theories had developed on the basis of the concept of waves rather than particles. Then we would probably talk not about electrons, semiconductors, or silicon but about indices of resistance and patterns of refraction. But there would still be a limit on the speed with which messages could be conveyed by means of silicon materials. If both sets of representations were available, one could demonstrate that the limit expressed through one representation was isomorphic with the limit expressed in the other.

Note that I am not saying constraints tell us what reality is. This they cannot do. But they can tell us which representations are consistent with reality and which are not. By enabling this distinction, constraints play an extremely significant role in scientific research, especially when the representations presented for disconfirmation are constrained so strongly that only one is possible. The art of scientific experimentation consists largely of arranging situations so that the relevant constraints operate in this fashion. No doubt there are always other representations, unknown and perhaps for us unimaginable, that are also consistent with reality. The representations we present for disproof are limited by what we can imagine, which is to say by the prevailing modes of representation within our culture. But within this range, invariable physical constraints can operate to select some as consistent with reality, others as not. The search for knowledge in the theater of representation thus proceeds by nay-saying rather than yea-saying. We cannot see reality in its positivity; we can only feel it through universal constraints operating upon local representations.

The representations that many scientists now find interesting are self-similar with the contemporary episteme, in the sense that they are themselves models of the aleatory and ambiguous connection between representation and reality. Focusing on complex systems that are inherently unpredictable, chaos theorists recognize that chance variations are intrinsic to these systems, and consequently that their representations will never coincide with the system's actual behavior. To this extent, chaos theorists are closer to contemporary critical theory than to Euclid, Newton, or Einstein. Yet there remain significant differences in emphasis between contemporary literature and science. To elucidate them, I turn now to the articulation of local knowledge within critical theory.

Local knowledge has become such an article of faith in the human sciences that it is paradoxically on the verge of becoming a universal in its own right. The transformation of local knowledge into global theory occurs as follows. First, the theorist insists that any attempt to create a global theory will be defeated by the intransigence of local sites that refuse to be co-opted into the global. Then he takes the further step of recognizing that this vision itself constitutes a global theory. Since it cannot be refuted through locality, it is the most successful and complete global theory possible. There is a growing inclination within literary circles to regard deconstruction in these terms, as a theory of local knowledge more totalizing than the totalizing theories it criticizes.[8] Obviously, most of the people who make this observation are not deconstuctionists. In Paul de Man's seminal article "The Resistance to Theory" (1982), a similar argument is made by someone on the inside rather than the outside. The result is an extremely complex dialectic in which critic and discipline, local site and global theory, rhetoric and grammar, theory and resistance to theory become so intertwined that each simultaneously implies and repudiates the other.

The opposition that de Man places at the center of his text is the distinction between grammar and rhetoric. The essence of grammar, de Man says, is its universality. Grammar, like logic, must apply to many different texts and language usages, for if it did not, it would not have the force of rule we associate with it. Rhetoric, by contrast,

[8]See, for example, Wendell V. Harris, "Toward an Ecological Criticism: Contextual versus Unconditional Literary Theory," *College English* 48 (February 1986): 116–131.

is unique to a given text. Tropes, in particular, "pertain primordially to language" and thus cannot be excavated from the language that constitutes them. In the act of reading, the "latent tension between rhetoric and grammar precipitates out," for grammar tries to account for rhetorical figures through extralinguistic generalizations that aim toward "mastering" the text. The attempts of grammatical readings to arrive at total accounts of the text are destined to be unsuccessful, however, for "*there are elements in all texts* that are by no means ungrammatical, but whose semantic function is not grammatically definable" (pp. 15–16; emphasis added).

Let us pause to think about this claim, for it acts as a pivot in de Man's argument. How does one prove that grammatically undecidable elements exist *in all texts*? Clearly it is not enough to give multiple examples, as de Man does so persuasively in *Allegories of Reading* (1979a), for the skeptic can always respond by saying that although the claim may be true for text *A*, text *B*, text *C*, . . . , that does not mean it is so for all texts. This objection holds even if the examples number in the thousands, for as long as one text remains unexamined, it is possible that the generalization may be false. In fact, to make the claim in this strong universal form, one cannot use an empirical method. The only way to prove that the claim is true *for all texts* is to use a method analogous to a mathematical proof, that is, to limit oneself to the rules of logic and the assumptions implicit in the proposition in order to prove that the conclusion is irrefutable. Of course de Man does no such thing, for the good reason that no one could do it. Propositions of this kind are not susceptible to proof in the mathematical sense.

Here you may object that I am putting too much weight on a careless overstatement, a momentary slip of the pen. Were the article to end with the claim that grammar will always be defeated by the resistance of undecidable rhetorical elements, the objection would be well taken. The claim would be worth making even if it were true only in a certain number of cases and hence would not depend on universality for its force. The rest of the article, however, makes clear that the claim's universality is at the heart of what de Man means by "resistance to theory" and thus is central to the argument as a whole. De Man wants rhetoric not only to resist grammar but to resist it irresistibly. Rhetoric can unequivocally defeat totalizing grammars only if it is *always* able to resist; and this ability requires that rhetoric, symbol and figure of resistance, harbor within itself a generalization, namely, the universal claim at issue.

Hence de Man concludes that

> technically correct rhetorical readings may be boring, monotonous, predictable and unpleasant, *but they are irrefutable.* They are also totalizing (and potentially totalitarian) for since the structures and functions they expose do not lead to the knowledge of an entity (such as language) but are an unreliable process of knowledge production that prevents all entities, including linguistic entities, from coming into discourse as such, they are indeed universals, consistently defective models of language's impossibility to be a model language. [De Man, 1982: 20; emphasis added]

For de Man, to recognize the universality of rhetoric's defeat of grammar is to recognize that rhetoric as a theory of reading is able to resist every universalizing claim except its own universality. The more successful the theory is, the more its insistence on locality is transformed into a kind of universality, so that "the language it speaks is the language of self-resistance." True to his theory, de Man ends by emphasizing that not all locales can be subsumed even into theory's resistance to itself, for "what remains impossible to decide is whether this flourishing is a triumph or a fall" (p. 20). To end with such a poignant and paradoxical interpenetration of the local and global shows what a consummate master of rhetoric de Man was.

But wait. How did we arrive at this point again? Through the claim that there are rhetorical elements *in all texts* which resist grammatical coding. And where did this claim come from? Not from any method of proof that would force us to the "ineluctable necessity" of recognizing its truth (p. 17), but from the assertion of the writer. The necessity, in other words, must be located within the writer's psychology rather than within the nature of the case itself. What makes this conclusion so curious is that at the beginning of the article de Man writes with great scorn about the psychological aspects of teaching. "Overfacile opinion notwithstanding, teaching is not primarily an intersubjective relationship between people but a cognitive process in which self and other are only tangentially and contiguously involved. The only teaching worthy of the name is scholarly, not personal" (p. 3). De Man makes clear that for him scholarship is an escape from psychology, not an exhibition of it. As in the return of the repressed, psychology nevertheless comes back, for it is only through psychology that we can understand the springs of that "ineluctable necessity" to forge a global theory of local knowledge. Knowing, then, that if de Man were alive he would detest the argument I am about to make, I will make it anyway in the hope of understanding not just personal quirks in his writing but larger patterns of thought characteristic of the ideology of local knowledge.[9]

The article is so cunningly structured that, to do it justice, I will need to look closely at how it reaches the crux of locating the resistance to theory within theory itself. In the opening paragraphs, de Man sketches in the following context. He had been asked to write the article on theory for the Modern Language Association volume *Introduction to Scholarship in Modern Languages and Literatures.* Clearly what the MLA committee had in mind was a survey of the field which would identify major trends and lay out the important issues. Instead, de Man wrote an article (presumably the one we are reading, or a draft of it) which valorized only one theory—his version of rhetorical reading—and characterized other kinds of theory as either nontheory or muddled resistance to theory. It is scarcely surprising that the MLA committee declined to use the article in their volume. De Man says he "thought their decision altogether justified, as well as interesting for its implications for the teaching of literature" (p. 3). What are those implications? For de Man, the teachable is the transmissible, and the transmissible is the generalizable. But if the essence of literariness is the nongeneralizable, "then scholarship and teaching are no longer necessarily compatible" (p. 4). Immediately, then, the conflict between the local and global is set up as a central dialectic. At the beginning its focus is de Man versus the MLA committee, or more generally scholarship versus teaching,

[9]This chapter was written before de Man's reviews in *Le Soir* surfaced. The questions these reviews raise about de Man's inclination toward totalizing theories are no doubt relevant to my argument; but since my interest is in the turns that de Man's argument takes rather than his theories as such, I have chosen not to consider the reviews.

but it soon moves inward in tightening spirals until finally it locates the conflict within theory itself.

These spirals are constructed through a series of oppositional stances that are debates about what theory ought to do versus what theory is actually able to do. Theory ought "in theory" (p. 4) to begin by defining its object; but it is unable to do so because until recently, it has always relied on extrinsic systems ("philosophical, religious or ideological") for its internal organization (p. 5). Thus defeated, theory may as well start pragmatically. This observation leads de Man into a survey of New Criticism, which was a movement not because it was theory but because it was uniformly hostile to theory. What, then, is theory? "Literary theory can be said to come into being when the approach to literary texts is no longer based on non-linguistic, that is to say historical and aesthetic, considerations" (p. 7). Specifically, theory comes into existence with the advent of structural linguistics, for then it is based not on an extrinsic system but on the structure of language itself.

Locating the basis for literary theory within language is an ambiguous move, for while it "gives language considerable freedom from referential restraint," it also "makes [language] epistemologically highly suspect and volatile, since its use can no longer be said to be determined by considerations of truth and falsehood, good and evil, beauty and ugliness, or pleasure and pain" (p. 10). And in fact, this "autonomous potential of language," which de Man identifies with literariness, makes literary theory intensely threatening to some people. Those who raise objections to this view of language are, however, not worth answering, for their arguments are "always misinformed or based on crude misunderstandings" (p. 12). Thus dismissing as unworthy opponents those who resist "real theory" because they disagree with structuralist views of language, de Man locates the only meaningful resistance to theory within theory itself—and so arrives at the crux discussed above.

If we think about how this argument is constructed, we can see that it proceeds as a series of inward-turning movements, always in the hope of finding a position strong enough to be defended, an opponent worthy enough to act as a serious contender. Theory becomes real theory when it locates its basis within language; resistance becomes real resistance when theory resists itself; reading becomes real reading when it resists de Man's theory of reading. Finally we are left with the spectacle of de Man resisting de Man, for he can find an opponent powerful enough to be interesting only within his own constructs. He is certain that this is a flourishing, but whether it "is a triumph or a fall" remains undecidable. It is undecidable because at this point local knowledge is indistinguishable from global theory, just as de Man is indistinguishable from the activity of reading. As local site merges with global system, the psyche looks at its object but sees only another version of itself, in an endless mirroring of self-similarities.

Let us not forget, however, that this inward turning of the theory toward paradoxical self-similarity hinges upon the universal claim that is the crux of the argument. If the theory is not pushed to this limit—if one is content to say that some or even most texts are configured so as to have grammatically undecidable elements—then the paradox does not appear, for then one is simply saying that not all sites can be accounted for by the universal theory in question. Why does de Man want to push the local into the global? There may be a clue in the subtle irony with which he uses the word "mastery." "Mastery," in de Man's usage, connotes universalizing moves made by opponents to limit the scope and power of rhetorical theory. Stanley Fish is said to have written a "masterful" essay when he tries to "empty rhetoric of its epistemological impact" (p. 19); grammar aims toward "mastering" meaning when it tries to subsume rhetorical figures in grammatical codes (p. 16). When de Man creates a global theory of local knowledge, he simultaneously repudiates and practices mastery in this sense, for he resists totalization by totalizing. Mastery is intolerable because it is identified with totalitarianism; but it is also unavoidable, because the only way *always* to resist totalizing moves is through a theory more universally applicable than what is resisted. The ideology of local knowledge, pushed to the extreme, is thus inextricable from the totalitarian impulses it most opposes. The unflinching honesty with which de Man faces this paradox is admirable, for it implies a profound awareness that impulses toward mastery are still masterful even when they are directed against mastery. The dream of local knowledge, that it can always and everywhere defeat global theory, carries within it its own subversion.

Most theorists working with local knowledge do not push its claims as far as de Man does, and the paradoxes that are the crux of de Man's argument consequently do not appear. But the very excessiveness of his argument, in an article widely cited in the profession, indicates how much the current of our time runs toward local knowledge. Even in scientific disciplines with strong traditional orientations toward global theory, local knowledge has come sufficiently to the fore to make the local/global dialectic problematic in ways it was not previously. In humanistic fields, the bias toward local knowledge is stronger. Making global arguments now in literary criticism is akin to trying to write psychobiography during the heyday of the old New Criticism. No one actually prevents such arguments from being made; but the prevailing practices of the discourse community tend so strongly in contrary directions that anyone who attempts them feels an invisible pressure pushing against each paragraph, each sentence, each word. Surely it is significant that a thinker such as Michel Serres, whose mind seems naturally drawn toward the abstract and global, chooses to deploy his abstractions in favor of local knowledge.[10] When Serres explicitly entertains the question whether his work is local or global, he invariably represents it as local although it partakes at least as much of global abstraction. A similar point could be made about Foucault's representation of his work in comparison with the work itself.

What in the present cultural moment makes the case for local knowledge seem so compelling? Why has it come to the foreground in such diverse projects as the New Historicism and nonlinear dynamics, fractal geometry and social archaeology, rhetorical criticism and strange attractors? What does it signify for our culture, our discourses, and our future that we are currently much more inclined toward of diversity than homogeneity, particularity than generality,

[10]Maria Assad has convincingly argued that in Serres's later work, he turns increasingly toward the local, repudiating the impulse toward globalization which marked his earlier essays. See Maria Assad, "Michel Serres: In Search of a Tropography," in *Chaos and Order: Complex Dynamics in Literature and Science*, ed. N. Katherine Hayles (Princeton University Press, forthcoming).

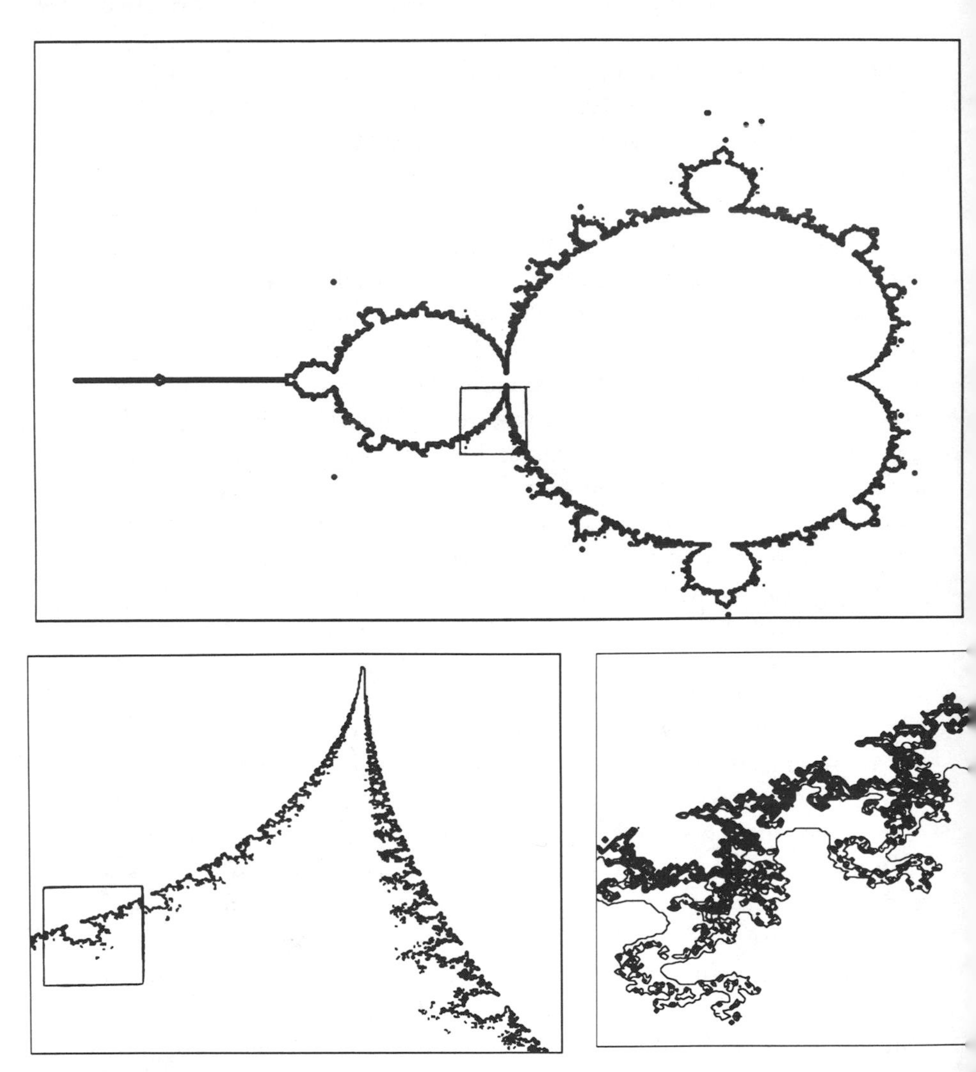

9. Mandelbrot set with progressively greater magnifications of boxed areas. Courtesy of Nural Akchurin.

refusal of mastery than mastery? The next chapters explore these questions in relation to Lessing's *Golden Notebook* and to postmodernism.

Meanwhile, from the information technology with which our future seems entwined are emerging images that I offer as correctives to the ideology of local knowledge. I am thinking of the haunting and evocative computer simulations of complex systems which show regions of overall symmetry intermixed with asymmetric hazy regions. When one of the hazy regions is magnified, it shows a mixture of symmetry and haze; and when one of the hazy regions of this magnified portion is magnified, it continues to show a mixture of symmetry and haze (figure 9). The image never resolves to either order or chaos, continuing to show infinitely fine-grained complexity for as long as magnification proceeds. Is not this an apt image for the interplay between the local and the global?

Laplace dreamed of an infinitely predictable world, but chaos intervened. De Man dreamed of a theory of representation that could not be universalized, but this in itself was a universal. Although global theory and local knowledge may each attempt to deny the other by extending its boundaries, when one is pushed to the limit, the other returns. Local knowlege and global theory are as much partners in the creation of representation as order and chaos are in the creation of the world. For as far as we can see or feel, clarity and haze, local variation and global form, intermingle.

CHAPTER 9

Fracturing Forms: Recuperation and Simulation in *The Golden Notebook*

Criticism on *The Golden Notebook* has largely been concerned to understand it as a work of literature, removed from its social context. Yet *The Golden Notebook* is deeply concerned with the relation between cultural formation, artistic expression, and individual identity. It foregrounds many of the concepts that are central to the new scientific paradigms—chaos and information, unpredictability and fragmentation—identifying them as crucial to the evolution of new forms of literature and subjectivity. Although Doris Lessing almost certainly did not know about chaos theory, she was keenly aware of the impact that proliferating information was having on traditional art forms. She also realized that these changes were bound up with the problematic relation of local sites to global perspectives. *The Golden Notebook* thus represents a development parallel to chaos theory. It testifies to how highly charged the idea of chaos had become by the mid–twentieth century.

The Golden Notebook (1962) represents a cusp in Doris Lessing's works, a transitional moment in which chaos is breaking open old forms but before it has catalyzed the emergence of "a new way of looking at life." And yet the fractured form points to the creation of something new—although whether this something is a novel remains problematic. Why a new form should be necessary is intimated by Anna as she writes about her dissatisfaction with her novel

The Frontiers of War. "The novel has become a function of the fragmented society, the fragmented consciousness. Human beings are so divided, are becoming more and more divided, *and more subdivided in themselves*, reflecting the world, that they reach out desperately, not knowing they do it, for information about other groups inside their own country, let alone about groups in other countries" (p. 61). To trace the connections between fragmentation and the changed novelistic function that Anna senses, it will be helpful to recall the distinction Walter Benjamin makes between the story, the novel, and information (1969:83–110).[1]

According to Benjamin, the story had its beginnings in craft labor. He conjures up an image of a storyteller spinning his tale while his listeners spin and weave cloth, with the rhythm of the story and the rhythm of the work interpenetrating. The purpose of the story is to give counsel, Benjamin says, which presupposes a physical community that speaker and listener cohabit, and more profoundly a community of values they share. The novel is distinguished from the story in being written. Following Lukács, Benjamin sees the novel as signifying a "transcendental homelessness" (p. 99), an essential isolation reflected in its attempt to grasp the meaning of life. Since this meaning can be apparent only in retrospect, after the whole shape of a life is known, the novel positions itself ontologically as a remembrance. The story, by contrast, addresses itself to how the listener can enrich his life as he lives it. Rather than try to grasp the whole, it contributes to the ongoingness of process.

The decay of community implicit in the contrast between story and novel progresses further with information. Unlike either story or novel, information has no depth, no spaces where the listener or reader may enter to supply her own interpretation. On the contrary, Benjamin says, the primary requirement for information is that it be "'understandable in itself'" (p. 89), fully explained and immediately accessible. Whereas story presupposes a shared context and commu-

[1]I am indebted to Alan Swingewood for drawing the connection between Benjamin's theory and Lessing in his article "Structure and Ideology in the Novels of Doris Lessing" *Sociology Review Monograph: The Sociology of Literature* 26 (1978): 38–54. Swingewood also notes that the emergence of a novel of information is connected with the disappearance of the "problematic hero" in Lessing's fiction and with her replacement by a "collective hero and conceptual structure" (p. 41).

nal values, information can be transported into any context. The incongruous juxtapositions of newspaper fillers illustrate the disappearance of shared, stable contexts: how many camels were born last year in Saudi Arabia, how many months an elephant carries its young before parturition. What connection does this information have to the reader, and what contexts does it presuppose and create? According to Benjamin, the triumph of information over story and novel bespeaks the death of experience, for to him experience means something more than facts or explanations. It signifies the texture of life created from a wealth of small shared moments, the unspoken and unconscious expectations that form the fabric upon which the figures of story are embroidered. In the absence of these shared contexts, experience is incommunicable. Facts can be told; events can be narrated; but experience cannot be shared, for the cloth out of which it is woven has been destroyed.

The Golden Notebook records the end of story through Anna's failure to tell a story, linking it to a fragmentation so profound that even the remembrance of experience is lost. Throughout, Anna senses that words are losing their meaning; "the gap between what they are supposed to mean, and what they in fact say seems unbridgable" (p. 300). The retrospective shape of the novel haunts her as she tries to write about the breakup of her five-year relationship with her married lover, Michael, using the characters Ella and Paul. Musing on how the ending of the affair distorts the telling of it, Anna writes that the "trouble with this story is that it is written in terms of analysis of the laws of dissolution of the relationship between Paul and Ella. I don't see any other way to write it. As soon as one has lived through something, it falls into a pattern. And the pattern . . . is untrue. Because while living through something one doesn't think like that at all." Anna wonders if she should try to tell her story by recording every detail of two days, one near the beginning and another near the end, to convey the sense of process. But she realizes that "I would still be instinctively isolating and emphasizing the factors that destroyed the affair. It is that which would give the thing its shape. Otherwise it would be chaos. . . . Literature is analysis after the event" (pp. 227–228).

Dismayed by "the thinning of language against the density of our experience" (p. 302), Anna turns to visual media as truer modes of expression, imagining the stories she tells as films rather than novels or stories. But even films desert her when she dreams that *The Frontiers of War* has finally been made into a movie. She complains to the director that the film has falsified her experiences. More than that, it has displaced them, for after seeing the film she finds that she can no longer remember how Maryrose held her head or Paul smiled (pp. 524–525). Later, the film moves inward and an imagined "projectionist" screens it within her mind. Thinking how wrong it is, Anna is devastated when she reads the credits and finds that *she* is the director. The problem thus goes deeper than language, deeper even than sensory perception. At its root is a fragmentation of the social context reflected in the four notebooks Anna keeps—black for her memories of her African experiences, red for her encounters with the British Communist Party, yellow for the stories she makes from her experiences, blue for her record of the experiences themselves. The "Free Women" sections do not, as they may at first seem to do, succeed in creating a metacontext into which all the other contexts may fit. Rather, they represent yet another partial context—the traditional realistic novel that is no longer able to contain or express all of the fractured sites of Anna's consciousness. Increasingly Anna feels that she cannot capture the truth of her experiences in any of these modes. The failure of the film confirms that the gap between reality and representation is unbridgeable, for it implies that the person she was cannot communicate with the person she is, so that her experience is incommunicable even to herself.

Under this pressure, the function of the novel necessarily changes. If the experiences that constitute one's past cannot be shared even with oneself, there can be no question of representing through remembrance the shape of a life.[2] Within *The Golden Notebook*, a novel of sorts still exists in the "Free Women" sections, but it is punctuated by the notebook entries. Within the notebooks, novelistic fragments also exist. But as the narrative continues, stories increasingly give way to information. The black notebook ends as a

[2]Patrocinio P. Schweickart, in "Reading a Wordless Statement: The Structure of Doris Lessing's *The Golden Notebook*," *Modern Fiction Studies* 31 (1985): 263–279, also notices the problematic nature of the novels within the novel, calling *The Golden Notebook* "a collection of canceled novels, novels Lessing had refused to write but nevertheless had written" (p. 274).

collage of newspaper clippings. A similar fate befalls the red notebook. The yellow notebook fragments into pieces of stories spliced together into a pastiche. The blue notebook, which Anna dedicates to writing about events as they happen without making up stories about them, continues to record information from her fragmenting sensibility after the other notebooks have stopped. It too ends; yet its final page marks a turning. Having drawn heavy black lines across the pages of the other notebooks, Anna resolves to write no longer in separate texts, but to put "all of myself in one book" (p. 607).

This will toward synthesis drives her into a direct confrontation with the information age. She cuts out newspaper articles about horrific events all over the world—rapes, tortures, murders, genocide—and for as high as she can reach, papers the walls of her room with them. She spends her days going about the room, reading clippings and putting up more. By physically locating this disparate information within her living space and by absorbing it within herself, she succeeds in creating a common context for it. The price she pays is that she must now live in the space she has created. She is the common denominator for the information she scans; if her attention wanders or if she leaves the room, it degenerates again into fragments. "It was as if she, Anna, were a central point of awareness, being attacked by a million unco-ordinated facts, and the central point would disappear if she proved unable to weigh and balance the facts, take them all into account" (p. 649). Tortured by the information on her walls but unable to break free from her feeling that she must bring it together into a whole, Anna contemplates beginning to paper a second room with clippings. By extension, one could imagine a world in which the Derridean maxim that there is no outside to the text would be literally true. From her doubly interior position within the text/room, Anna realizes that her quest for wholeness (which is also Lessing's quest for a novel about Anna) has led her into chaos. She is surrounded by it; more than that, she has *become* it.

As an emblem, the text/room reveals that the attempt to see things whole entails recognition that the individual psyche is bound to the culture by feedback loops that run through textuality. The texts produced by the culture and by the individual, produce them in turn. If the texts of a culture are profoundly fragmented and obsessed with death, so is the text that is the individual. And if one wants to grasp the whole, as in a traditional novel, one must write about fragmentation and chaos, which is to say, one must write information and not a novel. This is the paradox at the center of *The Golden Notebook*, whose fractured form signifies that it is a novel under erasure, attempting to represent the shape of the whole by meditating upon all the complex reasons why wholeness is impossible at this time.

The Limits of Analogy: On the Cusp

In Lessing's 1971 Introduction to *The Golden Notebook*, she complains that the central importance of the novel's structure had been overlooked, asserting that it constitutes a "wordless statement" designed "to talk through the way it was shaped" (p. xiv). Since Lessing's Introduction, a number of essays have appeared emphasizing the significance of the structure and exploring its relation to the novel's central themes.[3] Several years ago Betsy Draine pointed out that Lessing's text, despite its concern with disorder, is highly symmetric (1980:32). Draine represented the order as follows (from left to right).[4]

[3]Barbara Bellow Watson, "Leaving the Safety of Myth: Doris Lessing's *The Golden Notebook*," in *Old Lines, New Forces*, ed. Robert K. Morris, pp. 12–37 (New York: Rutherford, 1976), comments on the tension between order and chaos in the structure, asserting that "nothing could be more orderly, nothing could be more a response to chaos" (p. 16). Herbert Marder, in "The Paradox of Form in *The Golden Notebook*," *Modern Fiction Studies* 26 (1980): 49–54, interprets the paradox of a formed formlessness as an embodiment of the mystical insights Anna has in "The Golden Notebook" and "Free Women 5." Anne M. Mulkeen, in "Twentieth-Century Realism: The 'Grid' Structure of *The Golden Notebook*," *Studies in the Novel* 4 (1972): 261–274, interprets the structure spatially, as a "crisscrossing of a multiplicity of viewpoints with a multiplicity of events and issues, near and far, through a series of stages in time" (p. 262). Schweickart, in "Reading a Wordless Statement," stresses the "articulated" nature of the structure, in the sense that it is "cracked, broken in pieces, and, at the same time, hinged, held together by folding-joints" (p. 268). Betsy Draine gives a subtle and complex reading of the form as typical of postmodernism. Draine sees the fractured form as a response to the "dynamic interplay of order and chaos" (1980:31), and she charts this interplay through the tension between nostalgia and irony within the narrative.

[4]The structure is actually more complex than this scheme recognizes, but this notation is useful for the moment. Later I will refine and modify it.

Free Women 1	Black Notebook 1	Red Notebook 1	Yellow Notebook 1	Blue Notebook 1
Free Women 2	Black Notebook 2	Red Notebook 2	Yellow Notebook 2	Blue Notebook 2
Free Women 3	Black Notebook 3	Red Notebook 3	Yellow Notebook 3	Blue Notebook 3
Free Women 4	Black Notebook 4	Red Notebook 4	Yellow Notebook 4	Blue Notebook 4

The Golden Notebook

Free Women 5

It is immediately obvious that the structure is recursive, cycling through each of the "Free Women" sections and four notebooks in turn, until the symmetry is broken with the part named after the whole, "The Golden Notebook." Just as there are four notebooks, so there are four recursions before this bifurcation is reached. Given these symmetries, it is tempting to think of the structure as bringing order out of chaos. Such a reading would argue that *The Golden Notebook* functions like a self-organizing system, with the onset of chaos precipitating a reorganization of the text at a higher level of complexity.

In my view this reading ought to be resisted. The analogy with chaos theory is compelling. It is also subtly misleading if it leads us to read *The Golden Notebook* retrospectively. From a retrospective view, with knowledge of the paths that were in fact taken in the development of chaotics, earlier works that embody some but not all of its features are seen as imperfect avatars pointing toward later developments, as though there were some inevitable and necessary progressions these events must follow. For writers struggling to conceptualize new ideas before they crystallized into paradigms, however, it would not be clear that the situation consisted of developments, that these developments could be placed along a line, or that this line was a path pointing in a new direction. As a text located at or near a transformation in paradigms, *The Golden Notebook* almost certainly contains possibilities that would be suppressed if it were fixed through coordinates that became points of reference only after it was written.

In juxtaposing *The Golden Notebook* with chaos theory, then, I do not intend to argue that it embodies a recursive structure in the same sense chaos models do. One of the major contributions of chaos theory has been to provide a set of ideas that people can use to think with when they confront the apparent paradox that chaos makes order possible. Lessing did not have access to these ideas when she wrote *The Golden Notebook*. Consequently, the problems involved in wresting order from chaos seem to her more intractable and less susceptible to rational analysis than they would to someone whose thought has been formed by chaos theory. The recursive structure of *The Golden Notebook* does share certain general characteristics with chaos models, including the tendency to magnify latent instabilities and to constitute the relation between local sites and global perspectives as problematic rather than self-evident. But Lessing's text differs significantly from the sciences of chaos in its vision of how order and disorder relate to each other.

This complex relation is suggested by two complementary dreams in *The Golden Notebook* about spatial fragmentation and unity. In the first, Ella dreams of the house in which Paul lived with his wife, Muriel. Ella imagines that *she* now lives in "the ugly little house, with its little rooms that were all different from each other." She attempts to unify the house by redecorating, but finds that "it was holding together precisely because each room belonged to a different epoch, a different spirit" (p. 225). The dream inscribes the belief that fragmentation is necessary to prevent a collapse into chaos. At the same time, the fractured contexts of the rooms make coherence impossible. Fragmentation is thus both a symptom of chaos and a defense against it.

Later Anna dreams that she is looking down at the globe covered by a "web of beautiful fabric stretched out." The fabric glows with colors representing the various countries—red for the Soviet Union, black for Africa, and so on. As she watches, the colors melt and flow into each other, forming "one beautiful glittering colour, but a colour I have never seen in life." Feeling "unbearable happiness," she realizes that the "world was slowly dissolving, disintegrating and flying off into fragments, all through space, so that all around me were weightless fragments drifting about. . . . The world had gone, and there was chaos. I was alone in chaos." Unable to catch the meaning of the dream before it vanishes, she remembers only the phrase "somebody pulled a thread of the fabric and it all dissolved" (p. 299). The dream suggests that wholeness and chaos are very close; all that stands between them is a loose thread. Unity is thus both the opposite of chaos and the prelude to it.

Taken together, the two dreams make a larger statement. Ella's dream expresses the awareness that what is taken for global order in classical paradigms is in fact only the partial order that each room possesses when considered by itself. As long as one is content to stay within a single room (each belonging, Ella says, to a "different epoch, a different spirit"), one can console oneself with the generalizations that Lessing, in her essays written during this period, consistently labels "parochial."[5] But when one attempts a transcendent perspective such as Anna achieves for a moment in her dream, it is unclear whether the larger picture will cohere into a mystical unity or disintegrate into dissolution and chaos. The effect is very much like what Anna feels in her dream, a complex undecidability that hovers just outside the horizon of comprehensibility.

This effect is quite different than that conveyed either by chaos theory or by deconstruction, each of which is (like Lessing's text) centrally concerned with disorder. Although some "fuzzy" regions always remain within maps of chaotic systems, chaos theory provides a model for the mingling of order and disorder which is entirely rational and deterministic. As we saw earlier, the commitment to rationality and to a noncontradictory global theory is one of the important differences between chaos theory and deconstruction. Deconstruction, for its part, has renounced the possibility embodied in Anna's dream, that a transcendent viewpoint from which the whole makes sense can be achieved. In *The Golden Notebook*, the tension between order and disorder has been neither rationalized, as it is in chaos theory, nor sharpened into an aporia, as it is in deconstruction. Rather, chaos and order mirror each other, making the text's stance toward them undecidable in a way that is distinctly its own. Later I shall return to explore the implications of this reflexivity for

[5]See, for example, Lessing, 1974:17.

the peculiar ending of Lessing's text. For the moment, I want to emphasize that because *The Golden Notebook* is on a cusp, it can keep in suspension possibilities that later will be forced into one path or another when the new paradigms become fully articulated.

Disjunctions and Recursions: Global Visions and Local Politics

The two dreams related above are characteristic of Ella and Anna, in that they show Ella's concern with the local and domestic, Anna's with the global and public. Ella is Anna's fictional creation (just as Anna is Lessing's), and her anxieties are typically weaker, more conventional versions of Anna's concerns. Anna is worried about global destruction, Ella about clashing wallpaper. Like Ella, Anna is caught in a love triangle with her married lover and his wife; but she is never reduced so completely to the local as Ella is. Running alongside her account of her affair with Michael are her meditations on her political activities in Africa and her experiences with the British Communist Party. However, the distinction between Anna and Ella is not merely that Ella is Anna writ small. Part of Anna's problem is that she cannot see how to connect the domestic and public, the local and global. In separating off a part of herself into Ella, Anna practices fragmentation as a strategy of containment against chaos, just as Ella does in her dream.

The disjunction between the local and global is one of Lessing's major themes, appearing in different guises in the various notebooks. In the blue notebook, it manifests itself in Anna's belief that she fell in love with Michael because he was a "real man," calling forth all of her response as a "real woman." For Anna, the proof of these essentialist ideas is the vaginal orgasm, which she stoutly defends as an infallible sign that the love she feels is real. Yet on September 17, 1954, the day she resolves to write everything down just as it happens, she records that Michael awakes and takes her sexually because he hears Janet moving in the next room. She tries to reason away her resentment at Michael's jealousy of her child, asserting that "the resentment, the anger, is impersonal. It is the disease of women in our time" (p. 333). Nevertheless, when they are finished, Anna feels resentment that is "suddenly so strong I clench my teeth against it" (p. 334). She tries to make the best of the situation, telling herself that the "two personalities—Janet's mother, Michael's mistress, are happier separated. It is a strain having to be both at once" (p. 336). Ironically, this separation ensures that even if Michael were willing to leave his wife (and he is not), Anna and her child could not live an integrated life with him. Living whole would require Anna to confront Michael with her anger. She understands that if this confrontation were to occur, it could not be confined to the personal level. Behind her anger loom the assumptions of a sexist society, just as behind his patronizing tone lies the tortured history of mid-twentieth-century Europe, including a family killed in Nazi death camps and friends murdered in Prague by Communists. Paradoxically, fragmentation is necessary because everything is connected.

Anna's problems with Michael are rendered more problematic and intense by the fractured contexts in which she lives. When Anna refuses to take her anger seriously because "it is the disease of women of our time," she is caught in the disjunction between her immediate reality and her global analysis. Avoiding confrontation, she reassures herself that her relationship with Michael has a core of reality because the stereotypical responses she uses to constitute it are real. However, these responses reinscribe the very cultural history that she seeks to escape when she insists they are genuine and spontaneous expressions of her deepest self. To prevent these contradictions from surfacing, she uses suppression and fragmentation. Refusing to think about what Michael's alienating and alienated comments signify, she finds her deepest pleasure in cooking for him, in part because cooking does not engage her critical analytic self.

The problem with this reading is that it makes it sound as if Lessing intended to expose the contradictions Anna does not see. Ellen Morgan, in her excellent article "Alienation of the Woman Writer in *The Golden Notebook*" (1973), has shown that Anna's refusal to criticize Michael's jealousy is part of a pervasive pattern in *The Golden Notebook* in which women negate their own responses when they are critical of men. She draws the almost unavoidable inference that such attitudes are also Lessing's (p. 480). The consistency with which double standards for men and women appear in

The Golden Notebook and elsewhere in Lessing's writing has led Marion Vlastos Libby to say plainly, "Doris Lessing is not a feminist, and *The Golden Notebook* is not a feminist novel" (1974:106). I agree with these conclusions; as Morgan and Libby argue, there is every indication that Lessing sympathizes with Anna's viewpoint. In her 1971 Introduction, for example, when Lessing scolded her readers for missing what she regarded as the point of her book, she set up feminist readings of the text *in opposition to* the themes of wholeness and fragmentation. Lessing's inclination to bracket off feminist issues, as if they were beside the point, suggests that she has less insight into the way gender construction relates to fragmenting contexts than the recursive symmetries of her text may suggest.

Yet she also creates women characters strong enough to be, as Libby says, "oddly at variance with the qualities revealed in their other functions in life—with their independence and strength and above all with their extraordinarily acute perception of the world" (p. 109). The recursive structure is an important factor in reconciling the tensions between Lessing's strong women characters and her antifeminist presuppositions, for it occludes as well as reveals correlations between different sites. In Ella's dream, fragmentation is identified as an obstacle to unified understanding. At the same time, the fragmented text allows different degrees of awareness to coexist, held as it were in suspension by the recursive structure. The result is a text fractal in its complexity, with areas of clear symmetry intermixed with other areas of suggestive but fuzzy replication. As the narrative scrolls through the various notebooks, the same problems and issues reappear at different sites. How clearly they come into focus at each site depends largely, however, on the modality of the site.

The red notebook, with its thrust toward global understanding and public action, has none of the obfuscations Anna practices in the bedroom. Here the relation between fractured contexts and the local/global split comes plainly into view. Anna says that she decided to join the British Communist Party because she felt the "need for wholeness, for an end to the split, divided, unsatisfactory way we all live" (p. 161). Yet joining the party "intensified the split" (p. 161), for in the wake of revelations about Stalin's regime, party gatherings splintered into factions. As attempts to overcome contradictions within the Party give rise to more contradictions, the atmosphere of party meetings edges toward hysteria. Anna recognizes the feeling from her recurring dream about the figure whom she calls joy-in-spite (p. 482). The dream figure, who appears in increasingly humanoid and familiar form as Anna comes closer to breakdown, is chaos incarnate. His appearance signals that, although fragmentation may initially be able to contain chaos, eventually it facilitates and accelerates the crackup toward which Anna and apparently the entire society are headed.

The yellow notebook manages to be at once more archetypal and trite than the blue notebook. The yellow notebook contains Anna's novel "The Shadow of the Third," which fictionalizes Anna's relationship with Michael in Ella's affair with Paul. The title refers only obliquely to Anna's relation to these two characters. Inside the novel the third is Paul's wife, Muriel, or more precisely Ella's projected image of Muriel. Claire Sprague, in "Doubletalk and Doubles Talk in *The Golden Notebook*" (1982), notes that patterns of twos and threes are pervasive in *The Golden Notebook*, constituting an elaborate play on doubles and projected selves. Aside from their psychological interest, these patterns are significant because they show unified contexts breaking up and re-forming into increasingly complex configurations. As two people are aligned against a third, then as a single person splits into two and one of these parts splits further to make a third, the fragmentation becomes extremely complex. It recalls the fracturing contexts of the British Communist Party, as well as the shifting triad of Molly, Anna, and Mrs. Marks, the psychiatrist whose name suggests that her Freudian generalizations are like Marxist analysis in their refusal to recognize the importance of local differences.

In the black notebook, the personal and public intermingle as Anna recalls how futile and contradictory were the efforts of the socialist group she joined in Africa. Dedicated to looking "at things whole and in relation to each other" (as Lessing remarks of Marxism in her Introduction, p. xiv), the group quickly split in two, then further splintered into factions within the subgroup headed by Willi, Anna's "official" partner. In retrospect, Anna can see how ironically ineffectual the group was. Following Marxist doctrine, they delared that the initiative for overcoming racism should come from the proletariat, which in this case meant local labor unions. Black labor unions did not exist, however, because blacks were not allowed to organize, and the white unions were even more racist than most of the society. Change, when it came, would be led by black nationalists. But the group disapproved in principle of nationalist movements, viewing them as a "right-wing deviation" (p. 90). Divorced from the realities of the situation, the group began to retreat from public action to private escape at the Mashopi Hotel.

As Anna moves deeper into her recollections, the black notebook becomes a kind of counternovel to her first novel, *The Frontiers of War*, which we know only through her parodic summaries and comments about it. In this counternovel, Anna recounts how the contradictions inherent in the group's situation exploded when Paul Blackenhurst, in a gesture that combined comradeship and arrogance, befriended Jackson, the Boothbys' black cook. As a result Jackson was fired, and he was forced to take a job where he would not be able to keep his family with him. When Anna expressed dismay, Willi pointed out that Jackson had held exceptional privileges and that most black men were not so lucky. "But I happen to know Jackson and his family," Anna objected, adding, "I can't believe what you say." Willi responded, "Of course you can't. Sentimentalists can never believe in anything but their own emotions" (p. 148). For Willi, only the global counts; the local does not matter.[6]

If the problem were merely a lack of connection between the local and the global, communicating links between them could perhaps be forged. The group's response to the African situation, for example, might have been corrected if they had started with local conditions and built up toward generalizations, rather than trying to impose a global theory onto the situation. One could imagine that they might have contacted the black nationalists and centered their efforts on helping them, rather than vainly looking to the labor unions. A similar argument could be made about Anna's personal relationships.

[6]This attitude is caricatured later by Molly's husband, Richard, who remarks that if he were the head of a secret police and given the task of identifying all the communists in the country, he would offer to set them up in a small health clinic in Africa. All the communists, he maintains, would say no. Although Richard has more than a few faults of his own, his analysis is acute enough to force Anna into an uncomfortable moment of recognition.

She judges her experiences with men according to a Platonic ideal that she holds in her head. If the relationship does not immediately conform to her idea of what it should be, she dismisses it as second-rate or not "real." The notion that one could *build* intimacy, that intimacy is the result of and not the precondition for shared experience, seems not to occur to her until she meets Saul Green, and then it takes a psychotic form. Like the socialist group in Africa, Anna rejects local reality because it does not conform to her global expectations.

Another version of the problem is to see people as merely symptomatic of their time, without individual personality. For example, Anna sees Nelson and his frenetic wife as the victims of fractured contexts that make wholeness impossible. She says to Mother Sugar that there is a "crack in that man's personality like a gap in a dam, and through that gap the future might pour in a different shape—terrible perhaps, or marvellous, but something new—" (p. 473). The perception is finely tuned. The scene with Nelson, his wife, and his friends at the party recalls Andy Warhol's quip that the essence of postmodernism is being hyper and bored. Through the surface of the boredom, crises keep erupting, fueled by the tremendous internal pressure created by fear of global destruction. The inscription of the global into these people overwhelms them as individuals, making it impossible for Anna to relate to them as anything other than symptoms of a common malaise. In a more conventional way, Ella senses the same thing about Cy Maitland. Yet both Anna and Ella remain confident that somewhere a "real man" exists who is so essentially himself, so sure of his identity as a man, that they will be able to respond to him as "real women" and thus be saved from being merely expressions of their time.

The problem, then, is not so simple as forging links between the local and global, for the global collapses into the local, just as the local reinscribes the global. In the last recursion through the blue notebook before the pattern is broken and a new kind of symmetry appears, Anna experiences a nostalgic glimpse of how the local/global problematic might be resolved if mediating links could be constructed. Returning to a childhood fantasy, she imagines herself playing "the game." First "I create the room I sat in, object by object, 'naming' everything, bed, chair, curtains, till it was whole in my mind, then move out of the room, creating the house, then out of the house, slowly creating the street, then rise into the air, looking down at London . . . but holding at the same time the room and the house and the street in my mind, and then England . . . then slowly, slowly, I could create the world." Having attained a position in space from which she could see the vast panorama of the cosmos, she would "try to imagine at the same time, a drop of water, swarming with life, or a green leaf. Sometimes I could reach what I wanted, a simultaneous knowledge of vastness and of smallness" (p. 548). It is significant that Anna's approach to this knowledge is through a truer naming, and that the naming comes out of a heightened individual sensitivity. The naming game implies that the local/global split could be resolved if she (and we) could break through to a purer use of language by practicing a more intense and aware subjectivity.

Subjectivity and Chaos: The Question of Form

In *The Golden Notebook*, the ordering principle that keeps chaos at bay is critical and analytical consciousness. Recall, for example, Anna's belief that her consciousness is the focal point that alone can provide the synthesizing perspective necessary to bring chaotic information into a semblance of order. However, the attempt by an individual consciousness to resolve the local/global split is complicated, because subjectivity itself is necessarily local and global, in the sense that it is at once intensely personal and an expression of its culture.

Subjectivity is a highly charged issue for Lessing. As a Western writer with communist sympathies, she is torn between locating the springs of art in an individual consciousness and locating them in a collective spirit. Anna, scanning material published in communist countries to find if it is suitable for British audiences, finds most of the writing "bad, dead, banal." She feels "forced to acknowledge that the flashes of genuine art are all out of deep, suddenly stark, undisguisable private emotion" (p. 349). She recalls a lecture on medieval art in which she told her audience that "art during the Middle Ages was communal, unindividual; it came out of a group consciousness. It was without the driving painful individuality of the art of the bourgeois era. And one day, we will leave behind the driving egotism of individual art. . . . Art from the West . . . becomes more and more a shriek of torment from souls recording pain" (p. 350). In the middle of the lecture, however, she started to stammer. "I know what that stammer meant," she declares, and thereafter gives no more lectures on art. Lessing quotes the passage in her 1971 Introduction, adding that Anna stammered because she was evading the recognition that "there was no way of *not* being intensely subjective; it was, if you like, the writer's task for that time" (p. xiii). On the one hand, Lessing implies that we believe we are self-determined individuals only because our culture makes us think so. On the other, we can understand our culture only if we plunge into the subjectivity it has inscribed within us.

In her Introduction, Lessing says that the way out of preoccupation with the "tiny individual who is at the same time caught up in such an explosion of terrible and marvellous possibilities, is to see him as a microcosm and in this way to break through the personal, the subjective, making the personal . . . into something much larger: growing up is after all only the understanding that one's unique and incredible experience is what everyone shares" (p. xiii). This resolution, like Anna's lecture, is evading something. If the individual is merely an expression of the culture, then the spark of "suddenly stark, undisguisable private emotion" that Anna identifies with "genuine art" has no meaning. A similar evasion occurs in Lessing's essay "The Small Personal Voice." There Lessing says she holds "the view that the realist novel, the realist story, is the highest form of prose writing" (1974:4). In pondering what it is that draws her to the great novels of the past, she admits that "I was not looking for a firm reaffirmation of the old ethical values, many of which I don't accept; I was not in search of the pleasures of familiarity. I was looking for the warmth, the compassion, the humanity, the love of people that . . . makes all these old novels a statement of faith in man himself" (p. 6). At the same time, "if one is going to be an architect, one must have a vision to build towards, and that vision must spring from the nature of the world we live in." (p. 7). How can the writer simultaneously express her culture and position herself above it, directing it into new paths? The conundrum suggests that the tension between the individual and the collective is not so easily resolved as Lessing asserts.

What Lessing circles around but never directly confronts is the possibility that the microcosm/macrocosm correlation between the individual and the culture may itself be an artifact of bourgeois individuality. As we saw in chapter 8, received ideas about Western individuality were transported into scientific and mathematical models during the Enlightenment through the assumption that complex objects could be represented as collections of autonomous units moving along a continuous time line. It is no accident that in these models one can move easily from the individual unit to the global system; such a movement is possible because the assumptions used to constitute the individual unit are replicated in the assumptions used to construct the system. In the new paradigms, movement up from the individual unit to the system, or down from the system to the individual unit, is no longer an assumed characteristic but a focus for inquiry. Lessing oversimplifies when she suggests that one can resolve this problematic simply by regarding the individual as a microcosm of the global system, for such movement is possible or easy only under a narrow set of conditions. More to the point, she reduces the tensions embodied in her text to a formulation far too neat to encompass their unruly complexities.

The Golden Notebook (as distinct from Lessing's representation of it) presents two different enactments of how individual sensibility and collective spirit relate. In the first, global vision and local experience are joined through heightened subjectivity. Because no fundamental changes are transacted in the way human consciousness is constituted or represented, this enactment defuses the more radical implications of a collapse into chaos. In the second enactment, the split between local and global mounts in intensity until it forces a crisis, from which emerges a text that may appear to be the same as the previous writing but in fact operates according to very different premises. These two enactments represent a profound ambivalence toward the transforming significance of chaos as a new paradigm. At issue is whether chaos can be contained within old forms of expression and representation, or whether it will precipitate a sea-change in human experience and art. In her later works, Lessing increasingly took the second path, welcoming and calling forth the transformative power of chaos. But in *The Golden Notebook* both possibilities are active; both call for belief and commitment. The result is a text that is enormously complex because it is balanced between two paradigms, partaking of both and reducible to neither.

This complexity is encoded within the novel's form. On the one hand, the form demonstrates the failure of the novel to cohere, consisting as it does of fragments spliced together. Yet the symmetry suggests that some kind of reorganization takes place in the inner "Golden Notebook," which ought then to be apparent in the final "Free Women" section. In this reading, *The Golden Notebook* as a whole would possess the coherence that eludes its fragmented parts, for it chronicles the emergence of a new mode of integrated experience and, presumably, a new way of writing. Whether we think that this reorganization has in fact occurred depends in part on who we think is writing the text.

As Joseph Hynes has pointed out in his meticulous study "The Construction of *The Golden Notebook*" (1973), the answer to "Who writes this text?" is not obvious. In one sense, of course, Lessing writes it; but within the text itself, Anna is the author. Which Anna? As Hynes observes, several versions of Anna are inscribed in the four notebooks and the "Free Women" sections. Moreover, the text we have before us is neither the notebooks themselves nor a transcription of them. Rather, it is an edited collage considerably more distanced than the original texts toward which it points. To explain who wrote the text as it is, Hynes suggests that we posit a figure whom he calls Anna-the-editor, a character who has already experienced everything the Annas of the notebooks and "Free Women" sections record, and who retrospectively assembles, arranges, and edits these accounts.

To illustrate how the edited text differs from the fragments that supposedly comprise it, consider how the black notebook starts. We are not given the beginning; instead, we are presented with a *description* of it. Anna-the-editor, writing in brackets, tells us that the "first book, the black notebook, began with doodlings, scattered musical symbols, treble signs that shifted into the sign £ and back again; then a complicated design of interlocking circles, *then words*:]" (p. 56, emphasis added). The semiotic level at which the black notebook began before it moved into language is thus suppressed, since in our edited version everything has been transcribed into words, even the nonverbal symbols. One way to understand this transposition is to suppose that it is a recuperation of the novel's traditional retrospective function that was apparently lost in the exploding fragmentation recorded within its parts. This recuperation in turn depends on the recovery of a subjectivity unified enough to engage in retrospection, that is, Anna-the-editor. In this reading, both Anna-the-editor and her completed text are wholes that contain but are not reducible to the sum of their parts.

This is not the whole story, however, for it does not explain the peculiarly *parodic* nature of the ending. Nor does it adequately recognize that the description of the notebooks that Anna-the-editor gives us is in many ways a diminished, attenuated version of what actually reading them would have been. The experience of moving through the notebooks, of encountering the transitions from one kind of handwriting to another, from handwriting to pasted clipping, from clear text to brackets, from brackets to canceled text, would surely have been different not just in degree but in kind from the information Anna-the-editor gives us. Thus as readers, we undergo the alienation that Benjamin identifies as characteristic of an information age, because we are reduced from experiencing the texts directly to being told about them by Anna-the-editor. This implies that the recuperation of a unified subjectivity and a retrospective novelistic function are not in fact recoveries of what had been lost, but a simulation that is essentially different from the original.

Break/Down/Up; or, What Happened at the Bifurcation Point?

In her Introduction, Lessing says that the plan of *The Golden Notebook* is to document the encroaching presence of chaos, until in the inner "Golden Notebook" even the traditional forms of time and space break down; out of this chaos something new is to emerge. The question that comes sharply into focus as breakdown approaches is whether individual consciousness, as it has been traditionally constituted in the Western tradition, can provide the unified context necessary to bring these splintering fragments together. Or more accurately, since what we are reading is a fictional representation of that consciousness, whether the conventions of realistic narratives can continue to constitute individual consciousness as a putative source for "genuine art," or whether the disintegration of the subject will precipitate a crisis in representation which makes a traditional novel impossible to write. The self-reflexive turn of this construction, which links the writing subject with the subject of the writing, becomes explicit in the inner "Golden Notebook," which records Anna's breakdown and her entry into new visions of space and time.

Even as Anna experiences herself fragmenting into many different people, a countermovement begins that emphasizes fusion and wholeness. The invisible "projectionist," running the film that is at once Anna's life and her writing, screens a part she has not seen before which shows Michael merging with his fictionalized double, Paul, to create a new person with "the heroic quality of a statue" (p. 618). This figure, himself an image of art joined with life, gives Anna a message that valorizes her earlier attempt to join the local with the global in the naming game. He tells her that he and she will never be great men or women, envisioning a new future for humankind. Rather, they are destined to be "boulder-pushers," using "all our energies, all our talents, into pushing that boulder another inch up the mountain. And they [the great men] will rely on us and they are right; and that is why we are not useless after all" (p. 618). The thought that she is a Sisyphean "boulder-pusher" remains with Anna throughout this section. It is echoed and reinforced in another playback of the film, when she recognizes for the first time how thoughtlessly cruel she has been to people around her. She realizes that "the film was now beyond my experience, beyond Ella's, beyond the notebooks, because there was a fusion; and instead of seeing separate scenes, people, faces, movements, glances, they were all together." She understands that the film is about "a small painful sort of courage which is at the the root of every life, because injustice and cruelty is at the root of life." She recognizes that she had concentrated on the "heroic or beautiful or the intelligent" because she could not accept this truth, and as a consequence missed seeing the central importance of "the small endurance that is bigger than anything" (pp. 635–636). In this new perception, scaling back her ambition does not signal failure. Rather, it represents a necessary acceptance of the local. It answers Maryrose's question, "What's wrong with not settling for second-best?" with the insight that to refuse the local is arrogance, not purity.

From the perspective of the "boulder-pushers" passage, readers who object to the conventional nature of the ending are clinging to the illusion of dramatic revolution rather than accepting the reality of incremental change. Throughout, there have been intimations that revolutions too often turn out to be merely changes in personnel which leave underlying power structures intact, like the prisoner and guard who change places in Anna's nightmare (p. 345). Real change, Lessing seems to suggest, comes only when one devotes one's efforts to improving (however slightly) the place where one is, rather than retreating to global visions of what the world should ideally be like.

And yet this is hardly the whole story. Certainly it is not the most interesting story. It elides the deeper questions raised by Anna's encounter with chaos, and provides no satisfactory account of the extraordinarily complex structure of the novel. If Anna-the-editor is reducible to a boulder-pusher, why would she need to create so incendiary a form and resurrect so many dangerous possibilities? In her Introduction, Lessing skips rather lightly over these issues. She asserts that "in the inner Golden Notebook things come together, the divisions have broken down, there is formlessness with the end of fragmentation—the triumph of the second theme, which is unity" (p. vii). Compare this description with the text we have; is it adequate? After the searing intensity of Anna's psychotic episodes with Saul, surely most readers will find the end of the inner "Golden Notebook" attentuated and laconic by comparison. The final "Free Women" section is even more of a disappointment, with Molly retreating into the bourgeois respectability of a marriage to a "progressive businessman" who resembles her ex-husband, Richard, and Anna announcing that she intends to work as a marriage counselor with Dr. North,[7] join the Labour party, and teach night classes twice a week for delinquent youngsters. For a character who has represented herself as having undergone radical personality transformations, including fragmenting into multiple personae and losing all sense of conventional time and space, these are mild decisions indeed. Rather than displaying increased complexity, the closing "Free Women" section seems to be a parodic recapitulation of the opening

[7]Dr. North's name seems to indicate that he is the model for the fictional Dr. West, for whom Ella works. If we may judge him by the portrait of Dr. West that Ella draws, Anna's decision to work with Dr. North to save troubled marriages seems even more questionable than it would otherwise appear.

"Free Women" chapter, demonstrating the complete futility of all that has happened between them.

In her 1971 Introduction, Lessing does not acknowledge this as a possibility. Instead she asserts that the inner "Golden Notebook" brings the fragmented sites of *The Golden Notebook* together into a successful synthesis, signified by the fact that it is written jointly by Saul and Anna. At this point, Lessing writes, "you can no longer distinguish between what is Saul and what is Anna, and between them and the other people in the book." According to her, this fusion of characters "is a way of self-healing, of the self's dismissing false dichotomies and divisions" (p. viii). Lessing's representation of the text here glosses over an important distinction. It is not quite true that Anna and Saul collaborate in writing the inner "Golden Notebook." Rather, they write the imagined precursor to the text we have, the Ur-text to which our text alludes when Anna-the-editor tells us, in brackets, "[Here Anna's handwriting ended, the golden notebook continued in Saul Green's handwriting, a short novel about the Algerian soldier]" (p. 642). A summary of the novel follows, *not the novel itself.* Thus Anna-the-editor writes the inner "Golden Notebook," just as she has written all that has come before and after.

Lessing's elision of Anna and Saul with Anna-the-editor is significant because it points to an undecidable crux within the text. If Anna-the-editor is really Anna-and-Saul, we have no way to know that. The undecidability is made more profound by the recuperation at the end. If a new order has emerged, its nature is such as to make it indistinguishable from the old. What does it signify that success looks so much like failure, and that the emergence of a new order cannot be separated from its suppression? These questions have entered the criticism on *The Golden Notebook* as a debate about whether or not the ending is satisfactory.[8] I want to recast the terms

[8]Readers who defend the ending include John L. Carey, in "Art and Reality in *The Golden Notebook*," *Contemporary Literature* 14 (1973): 437–456; Carey writes that the ending demonstrates that "to exist is better than not to exist, to struggle is better than to give in, to face the truth and live with it is the measure of an individual's maturity" (p. 454). Valerie Carnes, in "'Chaos, That's the Point': Art as Metaphor in Doris Lessing's *The Golden Notebook*," *World Literature Written in English* 15 (1976): 17–28, gives a somewhat more qualified endorsement, acknowledging that the ending is "tentative, modest, shot through with irony" (p. 27).

of that debate by suggesting that there is a deep ambiguity about the ending which eludes an either/or resolution. Rather than a new order emerging out of chaos, it appears that the old and new orders have interpenetrated in a way that eludes rational analysis.

To tell this story, it is necessary to put aside for the moment a recuperative reading and suppose that the ending is about a radical reorganization that emerges after all. How would individual consciousness be constituted, and through what narrative and literary conventions could it be represented? A reading that attempted to answer these questions would have to be centrally concerned with form, for if a sea-change is to take place, it could be signaled only by a transformation of the premises underlying representation. Let us return to the beginning, then, and cycle through the text one more time, now taking the other path at the bifurcation point.

The New Order: Emergence of a Simulacrum

In her talk with Molly at the beginning of *The Golden Notebook*, Anna says that a well-known painter has announced that he is giving up painting because "the world is so chaotic art is irrelevant" (p. 42). The statement recalls Morse Peckham's thesis in *Man's Rage for Chaos* (1967) that the purpose of art is to break paradigms and assault established conventions, so that the replicating mechanisms that ensure the stability of the species will be invested with the necessary flexibility to ensure that change can also happen. And yet art also creates form; indeed, one might say that the essence of an artistic creation is its status as a shaped artifact. An art object without form is a contradiction in terms. Even if its form is such as to convey an impression of formlessness, it has still been designed and shaped with this purpose in mind.

Throughout *The Golden Notebook*, Anna wrestles with the transformation that takes place when fleeting thoughts are transcribed

---

Typical of readings that find the ending disappointing is Kathleen McCormick, "What Happened to Anna Wulf: Naivety in *The Golden Notebook*," *Massachusetts Studies in English* 8 (1982): 56–62; McCormick says that the "reader is engaged, but ultimately disappointed by Anna's near-achievements" (p. 57).

into a text. Once on the page, the words seem more fixed than they ever did in consciousness—more *formed*, especially when they are thoughts about chaos. In trying to explain to Mrs. Marks why she does not write for publication any more, Anna refers to this transforming process as a falsification, insisting, "It's a question of form." She goes on to say that contemporary readers will tolerate or welcome almost any moral aberration in their reading; but what "they can't stand is to be told it all doesn't matter, they can't stand formlessness" (p. 474). In *The Golden Notebook*, Lessing confronts the possibility that formlessness is truer to this cultural moment than form. However, she has only the inherited forms of her cultural traditions with which to convey this vision. Richard's pious hypocrisy that "I preserve the forms" (p. 26) resonates until, near the end, it becomes Anna's explanation for why she will write *The Golden Notebook* as she does.

The moment when Anna conceives the text, and therefore the moment when Anna-the-editor is conceived, is represented within the text. It occurs when Saul gives Anna the first sentence of her novel, which is also the first sentence of *The Golden Notebook*. The structure is thus self-reflexive as well as recursive, in the sense that the parts (the Annas of the different sections, the transcriptions of the notebooks) are contained within the whole at the same time that the whole (Anna-the-editor, *The Golden Notebook* as a finished text) is mirrored within the part.

One of the characteristics of a self-swallowing structure is infinite regress. Recall Borges's meditation on *The 1001 Nights*.[9] Borges notes that enfolded among the tales that Scheherazad tells is the most dangerous story of all. It occurs on Night 602, the evening when she decides that she will tell her own story. One can imagine her starting the story; when she arrives at the point where she decides to tell her story, however, she slips from the original recitation into an inner story that starts the narrative anew. From this inner story, she will at the crucial moment of self-reflexive mirroring slip into a doubly interior story, and from there into a triply interior

[9]Jorge Luis Borges, "The Translators of *The 1001 Nights*," pp. 73–87, especially 86, and "Partial Enchantments of the *Quixote*," pp. 232–234, especially 234, both in Borges, 1981.

story. . . . Unless some outside force intervenes, she will be caught in an endless recursion, each story nested inside its predecessor like an infinite series of Chinese boxes. "Does the reader perceive the unlimited possibilities of that interpolation, the curious danger—that the Sultana may persist and the Sultan, transfixed, will hear forever the truncated story of *The 1001 Nights*, now infinite and circular," Borges asks (1981:234).

In *The Golden Notebook*, the moment of slippage is not so dangerous, for it is only the first sentence that is repeated, not the entire story. Coming at the end of the inner "Golden Notebook," however, this moment is enough to raise questions about the status of the "Free Women" section that follows. What if at the self-reflexive junction the narrative slips from the epistemological framework that contains the story up to that point and enters another frame in which the forms may be perserved but the entire orientation is different? In that case, the final "Free Women" section would be neither a realistic working out of scaled-down expectations nor a parody of the initial "Free Women" chapter, but a simulacrum, a copy that has no original. It has no original because it does not exactly repeat either the beginning or the penultimate inner "Golden Notebook." Nevertheless, it is a copy in the sense that it is caught in a recursive loop. As a result, the text is a representation of its failure to represent what it points toward but can never reach.

Anna says as much as she approaches the moment when the text will swallow itself like the uroborus, the snake with its tail in its mouth. "During these last weeks of craziness and timelessless," she writes, "I've had these moments of 'knowing' one after the other, yet there is no way of putting this sort of knowledge into words. Yet these moments have been so powerful, like the rapid illuminations of a dream that remain with one waking, that what I have learned will be a part of how I experience life until I die. Words. Words. I play with words, hoping that some combination, even a chance combination, will say what I want. . . ." If we take Anna at her word(s), whether the ending is a simulacrum or a representation is inherently undecidable, since the informing experience that distinguishes the two is itself beyond words. "The people who have been there, in the place where words, patterns, order, dissolve, will know what I mean and the others won't," Anna writes (pp. 633–634).

Even though the confirming experience cannot be shared through language, it effects a change that will make the illuminati cognizant that what follows is not the thing itself, or a representation of the thing, but a simulacrum. This knowledge registers itself as irony. "But once having been there there's a terrible irony, a terrible shrug of the shoulders, and it's not a question of fighting it, or disowning it, or of right or wrong, but simply knowing it is there, always. . . . All right, I know you are there, but we have to preserve the forms, don't we?" (p. 634). In this view, the conundrum of how to create a form that will represent formlessness, how to forge a context that will make sense of a context of no context, is to create a pastiche whose fragmentation accelerates until it implodes into a new kind of form, the simulacrum. This is the transformation that Anna senses in the depth of her madness, when she writes that something "has to be played out, some pattern has to be worked through" (p. 583).

To position this reading more fully, I want to compare and contrast it with Jean Baudrillard's (1981, 1983) concept of the simulacrum. To illustrate what Baudrillard means by a simulacrum, consider the following progression. In ancient times, animals were killed for food and their skins were used as warm, utilitarian coverings. As animals grew scarce, their furs became status symbols, signifying wealth and privilege. Since there were not enough real furs to go around, simulated furs were produced to fill the demand create by those who aspired to the illusion of wealth but did not possess the money to buy real furs. At first these synthetic furs were marketed as copies of the originals. As more and more copies were produced, however, they began to compete on the basis of their superiority to other copies. At some point the original disappeared altogether, no longer serving to anchor the chain of these proliferating signifiers. Then the copies "imploded" into a new order of nonreferential signification that operated by displacement rather than representation. Baudrillard calls this the "hyper-real," a theater where everything is at once nonreferential and as real as anything else.

The emergence of an information age was crucial in bringing about this implosion, for it provided the technology that distanced an object so far from its putative origin that the origin could not be recovered, even in theory. A capitalistic economy was also important, for the displacement of desire onto a succession of copies created the expanding commodity market that capitalism needs to stave off collapse. It goes without saying that Lacanian psychology and poststructural linguistics were also important precursors to the idea of the simulacrum, for they carried the absence of origin into the formation of the psyche and the structure of language.

To connect the simulacrum with Lessing's text, consider another progression. One of the earliest successful children's television shows was *Howdy Doody*. Howdy was technologically simple. He was an obvious if crude copy of a human being, and he needed a human nearby to manipulate him and supply him with his voice. Howdy was soon superseded by the cartoon programs he introduced, which were at a further remove technologically from their human inventors, and also at a further remove physically from the human form. With computer animation came figures that were increasingly less anthropomorphic, including characters that combined human and machine parts. Finally Max Headroom arrived on the scene. Max was not a copy of a human being, as Howdy was, but a simulacrum. Max's essential quality was his nonreferential status with respect to the human. Since he lived in the TV sceen, he was both a representation and the thing itself. This disturbing convergence of the representation with its referent is characteristic of the simulacrum. Once we have entered into this space, the distinction between the copy and the original ceases to have meaning. At this point we live within the hyper-real, and ourselves have become it.

In suggesting that the final "Free Women" section is a simulacrum, I mean to invoke the associations with which Baudrillard invests the term, for I think they are nascently present in Lessing's text. If Anna-the-editor is not a scaled-down version of the Annas who wrote the notebooks, she must be a simulacrum, a copy that has no original. Similarly, if *The Golden Notebook* is identical neither with its parts nor with a synthetic, integrated whole toward which it nostalgically gestures, then it too is a copy with no original. This reading brings out the most disturbing and refractory aspects of *The Golden Notebook*, for it places the text along a path that Baudrillard and others would develop further.

I do not want to push the analogy too far, because *The Golden Notebook* also differs from Baudrillard's simulacrum in important ways. For example, although the reality that the final "Free Women" section gestures toward cannot be represented, it may nevertheless exist, whereas Baudrillard's point is that reality and our representations of it have collapsed into the same space. Whether this implosion actually takes place in *The Golden Notebook* remains undecidable, an indication that Lessing's text is caught between two paradigms. It has no choice but to reenact the old, for those are the only forms available; and yet by pushing the forms to (and perhaps beyond) the limits of their assumptions, it gestures toward the new.

One of the insights that Anna has when she is listening to the "projectionist" is how thin the line is between her own writing and the "style of the most insipid coy woman's magazine"—perhaps the magazine Ella works for (p. 620). Changing "a word here and there only," Anna transforms her style into a parody of her writing and of herself. That thin line keeps *The Golden Notebook* from entering the hyper-real. In being able to distinguish her authentic voice from a parody, Anna retains a sense of the reality of her subjectivity, and consequently of its potential as a source for her art. Thus the ending can be read as a reinscription of the values that underlie the realistic novel, and more generally of the assumptions that make modernist representations possible. But it can also be read as signaling the transformation of the text into a postmodern collage of information, in which parody does not exist because the center did not hold. The ambiguity points toward a profound duality within the new paradigms—whether they imply the renewal of human subjectivity as it has traditionally been constituted or its demise.

CHAPTER 10

Conclusion: Chaos and Culture: Postmodernism(s) and the Denaturing of Experience

WHAT in the present cultural moment has energized chaos as an important concept? Why does it appear as a pivotal concept for us here and now? I conjecture that disorder has become a focal point for contemporary theories because it offers the possibility of escaping from what are increasingly perceived as coercive structures of order. But in privileging disorder, theorists cannot extract themselves from the weight of their disciplinary traditions, even if they want to (and scientists, for the most part, do not want to). Thus there arise complex layerings in which traces of old paradigms are embedded within new, resistances to mastery are enfolded with impulses toward mastery, totalizing moves are made in the service of local knowledge. The convoluted ambiquity that arises from these layerings is the leitmotif of this chapter; it is deeply characteristic of what I shall call cultural postmodernism. For it to come into being, earlier paradigms first had to be understood as *constructions* rather than statements of fact.

I define cultural postmodernism as the realization that what has always been thought of as the essential, unvarying components of human experience are not natural facts of life but social constructions.[1] We can think of this as a denaturing process. To denature

[1]I am grateful to David Porush for suggesting this metaphor to me and for providing valuable background material for this chapter in his book *The Soft Machine: Cybernetic Fiction* (New York: Methuen, 1985). Daniel Malamud helped me to unravel the complexities of denaturing as a technical process.

something is to deprive it of its natural qualities. Alcohol is said to be denatured, for example, when its natural chemical composition is altered by additives. Denaturing also carries the technical sense of altering macromolecules by treating them with chemicals or radiation. One of the molecules commonly treated in this way is DNA. Denaturing was an important step in breaking the secret of DNA's structure, for it allowed the double helix to be unwound. The unraveling of DNA's structure led in turn to bioengineering and to the possibility that human genetic material can be reconfigured. The denaturing process, then, is one of the technical developments that helped to constitute cultural postmodernism. It is also a metaphor for postmodernism's deeper implications. When the essential components of human experience are denatured, they are not merely revealed as constructions. The human subject who stands as the putative source of experience is also deconstructed and then reconstructed in ways that fundamentally alter what it means to be human. The postmodern anticipates and implies the posthuman.

The denaturing process proceeded as three waves of related events. Each wave was more or less distinct from the others, but they reinforced each other and at times flowed together. In the first wave *language* was denatured, in the sense that it was seen not as a mimetic representation of the world of objects but as a sign system generating significance internally through series of relational differences. In the second wave *context* was denatured when information technology severed the relationship between text and context by making it possible to embed any text in a context arbitrarily far removed from its point of origin. In the third wave *time* was denatured when it ceased to be seen as a given of human existence and became a construct that could be conceptualized in different ways. With language, context, and time all denatured, the next wave, as I have intimated, is the denaturing of the *human*. While this fourth wave has yet to crest, it is undeniably building in force and scope.

The Denaturing of Language

Of all the components of cultural postmodernism, the denaturing of language is the most extensively documented within the literary community, so I shall spend the least time on it. It resulted from the confluence of two currents of thought. One was expressed by attempts in the early years of this century to eliminate ambiguity and self-reference from formal systems. These attempts were sufficiently widespread to constitute an important part of the period's intellectual landscape. In mathematics, Whitehead and Russell led the way with the *Principia Mathematica*; in language theory, logical positivism held center stage with its program to purge discourse of imprecise utterance; and in physics, Einstein developed the special theory of relativity to provide an overarching framework within which observations from different inertial systems could be reconciled. These projects had in common the belief that it was possible to create a metalanguage that would not be contaminated by the assumptions of the object language. One by one, they were either discredited or reinterpreted to accommodate the realization that, as Niels Bohr observed, "we are suspended in language."[2]

The attempt to axiomatize mathematics led in 1931 to Gödel's theorem, which demonstrated that any formal system complicated enough for arithmetic must be either incompletely described or contradictory. Logical positivism, already in trouble by World War II, was dealt a death blow by such philosophers of science as T. S. Kuhn, N. R. Hanson, and P. Feyerabend, who argued convincingly that observational statements are always theory-laden. The special theory of relativity lost its epistemological clarity when it was combined with quantum mechanics to form quantum field theory. By mid-century, all three had been played out or had undergone substantial modification.

If we think of these projects as attempts to ground representation in a noncontingent metadiscourse, surely it is significant that the most important work on them appeared before World War I. Einstein published his papers on the special theory of relativity in 1905 and the general theory in 1916; the *Principia Mathematica* volumes appeared from 1910 to 1913; and logical positivism had its heyday in the closing decades of the nineteenth century. After World War I, when the rhetoric of glorious patriotism sounded very empty, it must have been much more difficult to think that language could have an absolute ground of meaning. Out of attempts to conceptualize symbols systems as unified fields of representation arose the real-

[2]For a fuller treatment of these developments, see Hayles, 1984.

ization that positing an all-inclusive field meant there was no vantage point outside the field from which to speak. Consequently, every utterance or representation already assumed what it purported to describe.

More or less independently, a second current contributing to the denaturing of language developed out of Saussurean linguistics. It is now an oft-told story that Saussure, by defining *la langue* as a sign system generating significance through differences between language elements, imagined language as an interactive field in which the meaning of any one element depends upon the interactions present in the field as a whole. Indeed, Saussure went so far as to assert that if any one word were removed—say, "pig"—the field would be altered, and consequently the significance of all remaining words would change. (To an Iowan, the assertion that removing "pig" would change the entire language is self-evident.)

Saussure's interactive view of language has certain affinities with the events sketched above, particularly with the development of quantum mechanics and relativity theory. There are also important differences. Whereas quantum field theory led to a very specific and limited sense of the uncertainty imposed by observing or speaking from within the field (defined by Planck's constant in the uncertainty principle), Saussurean linguistics was appropriated to support the deconstructive view that all utterance is ungrounded and indeterminate. These differences notwithstanding, quantum field theory has often been held up as validating the view that language is inherently self-referential and ungrounded. In this sense the two currents flowed into a single stream and constituted the first wave of cultural postmodernism, the denaturing of language.

By saying language was denatured, I mean to indicate a self-conscious attitude toward language in which signification is regarded as always already problematic. The sense that language is constantly unraveling, even as one weaves it into a design; that any utterance can be deconstructed to show that it already presupposes what it would say and hence has no prior ground on which to rest; that all texts are penetrated by infinite numbers of intertexts so that contextual horizons are always constructions rather than givens; in short, that signification is a construction rather than a natural result of speaking or writing—all this was implied by the denaturing process as it applies to language. Denatured language is language regarded as ground painted under our feet while we hang suspended in a void. We cannot dispense with the illusion of ground, because we need a place from which to speak. But it is bracketed by our knowledge that it is only a painting, not natural ground.

The Denaturing of Context

The denaturing of context crested after the denaturing of language, although both continue to be important components of cultural postmodernism. Just as World War I marked the end of the most ambitious and successful attempts to find a noncontingent ground for representation, World War II marked the point at which these matters ceased to be of merely theoretical concern and entered the mainstream of the culture through information technology. By World War II, transportation and aircraft had progressed sufficiently so that rapid troop movments were possible, which in turn made accurate and timely information as critical to the war effort as weapons and soldiers. World War II made information real.

The years immediately following the war saw an explosion of interest in information theory and technology. Research on information theory and systems was directly linked with the war effort. The cracking of the Enigma code by Alan Turing was crucial to Allied strategy in the last phases of the war; Turing was also responsible for proving certain theorems about information which greatly advanced the nascent art of computer technology. A similar connection exists in the work of Norbert Wiener. Although Wiener's contribution to the war effort doubtless loomed larger in his mind than in anyone else's, it was nevertheless his attempt to construct a guidance system for anti-aircraft weapons that directed the attention of theorists to information as a way to deal with intrinsically uncertain situations. Wiener, like Turing, was quick to see applications to postwar technology and promoted them in his book on cybernetics published a few years after the war.

Further evidence for the catalyzing effect of the war on information theory is Donald MacKay's account of how he began to formulate his ideas about information. "Towards the end of the Second

World War in the Admiralty's radar establishment," MacKay writes, "I found myself trying to follow the behavior of electrical impulses over extremely short intervals" (1969:1). MacKay, an important early researcher in the field, recalls that the idea of information was very much in the air during and immediately following the war. Additional indications of the importance of World War II are the dates of the first dramatic breakthroughs in information theory. Shannon's two seminal papers were published in 1948, Wiener's book on cybernetics the same year, Brillouin's analysis of information and entropy in 1951.[3]

Shannon's definition of information as a statistical measure of uncertainty had important implications for the relation of text to context. As long as meaning and information are connected, the information value of a message remains bound to its context. Consider, for example, how the meaning of "down" changes when I say it to my dog and when my stockbroker says it to me in response to a query about the market. If information is tied to meaning, its value is subject to change every time a message is embedded in a new context. Shannon, an electrical engineer working for Bell Laboratories, knew that information could not be used as the basis for a new technology unless it could be reliably quantified. From his point of view, an entity whose value changed whenever it was transported to a new place was a nightmare. He cut through the complexities of context dependence by the simple expedient of declaring that information had nothing to do with meaning. By implication, informational texts were also cut loose from their dependence on contexts.

At this point in my story, the first and second waves come together. Shannon, having cut information loose from its moorings in

[3]Shannon, 1948; Brillouin, 1951; Norbert Wiener, *Cybernetics; or Control and Communication in the Animal and the Machine* (Cambridge: MIT Press, 1948). The war may have left its stamp on information theory in other ways as well. We saw in chap. 2 that Shannon defined information as a measure of uncertainty over the strenuous objections of other theorists such as Brillouin. Was it the war that made the association of information with uncertainty seem compelling to Shannon and to later theorists who adopted his convention? Did the anxiety of trying to interpret uncertain wartime information facilitate the disjunction of information from meaning? Mary Lou Emery, in a private communication, has observed that these questions are importantly related to the way gender formations were reconstituted in the aftermath of World War II.

referentiality, was forced to define it internally through relational differences with other elements in the appropriate symbol set. Thus Shannon's theory of information is conceptually analogous to Saussure's theory of *la langue*, as well as to certain aspects of quantum field theory. But Shannon's move came at a decisively different point in time, when cultural conditions were ripe for the rapid development of information technology. His move was important less because it was a theoretical innovation than because it initiated a feedback loop linking theory, culture, and technology.

The aspect of this loop that I shall concentrate on, and that constitutes the second wave of postmodernism, is the denaturing of context. Initially messages were separated from contexts because such a move was necessary to make information quantifiable. Once this assumption was used to formulate a theory of information, information technology developed very rapidly. And once this technology was in place, the disjunction between message and context which began as a theoretical premise became a cultural condition.

Among the first indications that separating messages from their contexts could have dramatic effects were the weapons-guidance systems that an improved information technology made possible. Primo Levi (1985), in his account of Jewish partisans fighting against the Germans in World War II, vividly recalls what it felt like for the weapons operator to be separated from the context in which the weapons were used.

> I was in the artillery, you know. It's not like having a rifle. You set up the piece, you aim, you fire, and you can't see a thing. . . . Who knows how many men have died at my hand? Maybe a thousand, maybe not even one. Your orders come by field telephone or radio, through earphones: left three, drop one, you obey, and that's the end of it. It's like bomber planes; or when you pour acid into an anthill to kill the ants: a hundred thousand ants die, and you don't feel anything, you aren't even aware of it. [1985:110–111).

Since World War II, weapons-guidance systems have of course become much more sopisticated, transforming the psychology of modern warfare. The awareness that our entire social context can be annihilated by weapons that we will not even see is no doubt one reason why contexts in general seem to us precarious, capable of instantaneous mutation or extinction.

Another area in which the separation of text from context has had dramatic effects is biogenetics. Although microbiologists would probably not say it this way, organisms are in effect viewed as informational texts that can be opened to a literal embodiment of intertextuality by a variety of gene-splicing techniques. Such techniques point to the deconstruction of the body as text, rendering problematic distinctions between originary texts and clones. Similar deconstructions are occurring with traditional ideas of parenthood as new birth technologies make it possible to withdraw eggs from a woman, freeze them for an indefinite period, fertilize them *in vitro*, and place them in the uterus of the same woman at a later time or in another woman. When the genetic text of the unborn child can be embedded in a biological site far removed from its origin, the intimate connection between child and womb which once provided a natural context for gestation has been denatured.

Test-tube babies are, of course, extraordinary instances that directly touch the lives of only a few people. But denatured contexts are not extraordinary. Far from being confined to the kind of events that make newspaper headlines, they are extremely common. Take MTV as an example. Turn it on. What do you see? Perhaps demon-like creatures dancing; then a cut to cows grazing in a meadow, in the midst of which a singer with blue hair suddenly appears; then another cut to cars engulfed in flames. In such videos, the images and medium collaborate to create a technological demonstration that any text can be embedded in any context. What are these videos telling us, if not that the disappearance of a stable, universal context *is the context* for postmodern culture?

G. W. S. Trow (1978), following a similar line of thought, has argued that contemporary Americans live "within the context of no context." According to Trow, context as such has disappeared because our lives are split between an enormous grid of two hundred million people and the intimate family circle gathered around the TV set. With very little in between, especially for the growing percentage of the population who live in large urban centers, these two very different communities of discourse try to pretend that they share the same context. Consequently, context becomes a construction rather than a natural result of shared activities.

One of Trow's best examples is the TV talk show. Talk shows

thrive on the audience's expectation that something indecorous is about to be said or done. But impropriety is possible only when a social context exists which can be violated. Since no shared context in fact exists, the host must create the illusion of one. (In rare cases in which the context is taken to be the studio, as on *Saturday Night Live*, the violations are of another order.) To make the double move of creating and violating context, the host arranges the guests so that they appear to form an intimate circle of friends. Each new guest then arrives on the scene as an intruder for whom space must be made. The host also tries to make the audience feel that it is part of this intimate social group, as by direct address. But the constant splicing into other contexts through commercials and station breaks reminds us that context *is always already a construction*, created for specific purposes and dispelled as soon as these purposes are achieved.[4]

Trow, an environmental and short story writer who publishes elegantly ironic pieces in *The New Yorker*, is not concerned with the academic controversies that swirl around postmodernism. Nevertheless, his observations about context illuminate much that has been written on postmodern aesthetics. Most theorists writing on postmodernism agree, for example, that contemporary architecture is an important site for displays of postmodernism. Why? In Trow's terms, because it provides an arena for exploring the problematics of decontextualization. Robert Venturi, in *Learning from Las Vegas* (1972), points out that the assumed context for the great gambling casinos and hotels is not the land on which they are built but the moving automobile from which they are seen. Similarly, Kenneth Frampton argues for "critical regionalism" because he is all too well aware that traditional connections between structures and their environmental contexts are up for grabs (Foster, 1983:16–30). If they were assumed, it would be redundant to argue that buildings should relate to their sites.

As contexts are increasingly seen as constructions rather than

[4]Perhaps the most blatant examples of created contexts are so-called media events, which exist only so that the media can film them. Trow considers such events cons. The way to recognize a con, he says, is to ask yourself, "Would this be happening if I were not here to see it?" If the answer is no, then *you* are the context for the event, and any other context is an illusion created to manipulate your responses.

givens, who controls which context for what purpose becomes an important question. Consider the term "context control," which entered the vernacular as a euphemism favored by government spokesmen. It implies that if one can control the context in which damaging information is released, one has a much better chance of controlling the way the information will be interpreted. It goes without saying that in these instances context is seen as a construction to be manipulated rather than a preexisting condition. Another example is the "disinformation" that government officials have acknowledged giving out to the domestic press as well as to foreign governments. Only in a (created?) context of national security is it plausible to distinguish between "disinformation" and lies.

Since its introduction by professional image makers, context control has become an increasingly widespread concern. It is a central issue in situations where computers are used to reconstruct incomplete or missing information. Satellite transmissions, for example, are customarily presented to the public, the media, and governmental decision makers in "image-enchanced" form: the original informational text has been fed into a computer and reconstituted into a more coherent image designed to enhance its salient features. The enhancement process is designed as an information-preserving transformation only for certain variables. Other variables are necessarily changed, since their alteration is what allows the image to be enhanced for the context of interest. The process blurs the boundaries between text and context, for an assumed context has *already changed the image before we see it*. In these instances controlling context is literally equivalent to controlling interpretation. The technological sophistication and flexibility of modern context control make it different in kind, as well as in degree, from the forms practiced in earlier eras. The Renaissance courtier who waited until his sovereign had dined before breaking bad news was practicing context control in rudimentary form. But only with the advent of modern information technology has context control ceased to be an occasional phenomenon and become a sophisticated set of strategies endemic in postmodern society.

So fluid and changeable are the contexts of contemporary life that new kinds of units, context-plus-text, are emerging. Max Headroom, for example, lived permanently within the context of the TV screen. Because they were inseparably fused, the framing screen and image provided the illusion of a stabilized context. But this new unit could in turn be embedded in other contexts, usually other TV sets, setting up a self-reflexive play between Max's TV screen and the TV screen framing it. The visual reflection of Max's screen in the surrounding screen is equivalent to the TV talk show's ploy of simultaneously creating and violating context. The stability that one screen bestows the other takes away. What better way to depict what it means to live within the context of no context?

From Denatured Contexts to Reconfigured Spaces

So thoroughly has context been denatured that it may be only a matter of time before the distinction between text and context collapses altogether. As early as the 1940s and 1950s, Borges was playing with this idea in such stories as "The Don Quixote of Pierre Menard" and "Tlön, Uqbar, Orbis Tertius" (1981). More recently, writers and theorists speak of contexts that cannot be characterized in conventional terms, as if their collapse had catapulted us into a new kind of space. In the science fiction novel *Neuromancer* (1984), for example, William Gibson imagines a world in which computer cowboys jack into microcomputers interfaced directly with their neural systems. When a cowboy fuses with a computer in this fashion, he enters cyberspace, a theater of operations beyond conventional time and space. In this new kind of space, intelligences (artificial and otherwise) sense each other's presences in ways unique to the medium. Cyberspace is a world unto itself, touching the conventional world at every point but remaining entirely distinct from it. The protagonist, a cowboy who has been altered by subtle chemical torture so that he can no longer enter cyberspace, considers life hardly worth living if he has to remain forever trapped in the ordinary dimensions of spacetime.

Gibson's vision has much in common with Baudrillard's theory of simulacra. Although Baudrillard (1981, 1983) tends to be more openly ironic about the social and technological forces that have caused late capitalism to "implode" and create a new kind of space, his concept of the hyper-real is like Gibson's cyberspace in that it images a space that touches the conventional world at every point but that has nevertheless undergone a radical transformation. As we saw in chapter 9, the hyper-real comes into existence when copies refer no longer to originals but to other copies; or more precisely, when it is impossible to distinguish any longer between a copy and an original. Like cyberspace, the hyper-real presupposes a radical erosion of context, for the sense that something is an original depends upon its association with a unique context. Consider: a London bridge is dismantled and transported stone by stone to the Arizona desert, where it is reassembled. The bridge is physically the same (within limits imposed by the plasticity of the material); but is it an original or a copy? The undecidability of the question illustrates how deeply the sense that something is a simulacrum is bound up with the loss of a stable context.

Both Gibson and Baudrillard are fascinated by the possibility that cryogenics may extend these kinds of questions to human beings. Is a Walt Disney frozen and brought back to life in the year 2050 the same man who created *Snow White*, or a copy? In *Neuromancer*, the corrupt and powerful clan of Tessier-Ashpool has for generations practiced cryogenics, so that its members have virtually all become simulacra (as their replicated names and numbers indicate). The prospect that human beings can become simulacra suggests that a new social context is emerging which will change not only what it means to be in the world but what it means to be human. Within this context-of-no-context, the postmodern shades into the posthuman.

Neuromancer is full of characters who have implanted into their bodies various kind of cybernetic devices, from the mirrored electronic eyes and retractable razor claws of the heroine to the shark skin and teeth grafted onto the face of a new-age punker. A common implant is a socket behind the ear into which computer chips can be plugged, creating a direct interface between human and computer memory. If one is going to an art auction, for example, one plugs in the appropriate chip and becomes an instant art expert. Such fusions of human organism and cybernetic mechanisms may seem far-fetched. But in fact cyborgs (a term coined from *cyb*ernetic *org*anism) already exist and are not particularly uncommon. About 10 percent of the U.S. population are cyborgs, including people who have electronic pacemakers, artificial joints, prosthetic limbs, and artificial skin.

As important as the increasing numbers of actual cyborgs are the ways in which we envision ourselves as parts of cybernetic circuits, even when our biological boundaries are intact. For example, Gibson in an interview (Greenland, 1986) described a teenager playing a video game as "a feedback loop of particles: the photons are coming out of the screen and going into the guy's eyes, and the neurons are moving through his body, and the electrons are moving through the computer." Physically intact, the player is nevertheless *already a cyborg*, for he is joined to the computer by a continuous interplay between his neural system and the computer's circuitry. In this view, to have nondetachable cybernetic implants is simply to reify the detachable connections that already bind humans to computers in thousands of video arcades and computer centers across the country. "I had a hunch from talking to people about computers," Gibson continues, "that everyone seemed to feel at some level, without really ever saying it, that there was space behind the screen. I just took that and ran with it as far as I could" (p. 7).

As the context-of-no-context transmutes into a new kind of space, the connections between denatured space and the new geometries become more apparent. In his influential article "Postmodernism, or The Cultural Logic of Late Capitalism" (1984), Fredric Jameson takes as an example of reconfigured space the Bonaventure Hotel in Los Angeles.[5] He points out that in contrast to the opulent entrances of older hotels, the Bonaventure's doors are obscure. Moreover, they open onto obscure spaces. He intimates that the Bonaventure in effect *disguises* its entrances because it "aspires to being a total space, a complete world, a kind of miniature city. . . . It does not wish to be a part of the city, but rather its equivalent and its replacement or substitute" (p. 81). Its public spaces are organized to make conventional reference points inapplicable. The lobby is surrounded by four identical towers, for example, so that the most prominent architec-

[5]The correct name of the hotel is the Bonaventura. For consistency, however, I will follow Jameson's usage in my discussion.

tural features are useless for spatial orientation. Streamers hanging over the four-story lobby make it impossible to estimate the lobby's volume and create a sense of a complex noncontinuous space that defeats usual lines of perspective. So far Jameson's description of the Bonaventure could, virtually without revision, be taken as an accurate description of cyberspace as Gibson represents it in *Neuromancer*.

When one reads about the feature that Jameson sees as confirming his diagnosis, the resemblance turns uncanny. The culminating sign of a reconfigured space, Jameson argues, is the Bonaventure's "great reflective glass skin," which he sees as analogous to "those reflector sunglasses which make it impossible for your interlocutor to see your own eyes" (p. 82). The parallel feature of *Neuromancer* is the heroine's mirror shades, which we only gradually realize are permanent cybernetic implants. In commenting on them, Brooks Landon has suggested that they represent a "polished chrome future in which reflection—surface—becomes the new reality," a future in which "electronic images, holographs, clones, cyborgs, and hallucinations intermingle . . . where almost everything reflects, replicates, or imitates something else" (1987:31). As Landon points out, and as the Bonaventure demonstrates, they are also a distinctive architectural feature of multinational corporate headquarters.

For Jameson, postmodernism is not a style but a "cultural dominant," emerging from capitalistic commodification carried to the extreme. As information replaces industrial production as the basis for the economy, capitalism does not disappear. Rather it enters its purest form, for now there is a quantifiable medium of exchange to which everything can be reduced, even human beings. Consequently, such high-tech buildings as the Bonaventure are not signifiers that point to technology as the signified. Instead, technology itself is a signifier, pointing to a mass of interconnected information networks of such enormous complexity that the human mind can no longer comprehend them. According to Jameson, that is why we structure our buildings and our narratives to image spaces so complex they elude human comprehension. Our technology is simply another image we make to body forth our sense of the global networks that control our lives, but that are so far beyond our powers of perception that we cannot even see them, much less control them.

The Denaturing of Time

What happens to time in postmodernism is very different from what happens to space. Whereas space detaches itself from mundane reality and forms a richly configured realm of its own, time sinks into the media experience of constructed, repetitious packages and becomes a series of disconnected intervals. As with much else, Borges anticipated this development in his essay "New Refutation of Time" (1964), which observes that human identity depends on memory and memory depends on seeing time as a continuous orderly progression. If time is cut loose from the idea of sequence, Borges suggests, every man who reads Shakespeare becomes, for that moment, Shakespeare. The cutting loose of time from sequence, and consequently from human identity, constitutes the third wave of postmodernism. Time still exists in cultural postmodernism, but it no longer functions as a continuum along which human action can meaningfully be plotted.

In *Being and Time* (1962), Heidegger used the finitude of human being as a fixed point from which to establish the possibility of an authentic life. Because we are finite, because there is an end point to our time as personally lived, it is possible for us to experience the dread that alone leads us to understand the nature of our condition. Derrida later argued (1976), in part on the basis of his reading of Heidegger, that there is no point of origin, that language and context are "always already" denatured. But still there was an end point to human being, and it provided the possibility of a stablized, if contingent, meaning. Since the 1960s, the consensus that there is a fixed end point has been eroded by our growing sense that the future is already used up before it arrives. Thus human being is left free-floating between an unimaginable origin displaced beyond itself before the beginning of time and an indeterminate end shifted into the present or even into the past.

How did we come to believe that the future, like the past, has already happened? Such terms as "postmodern" and "postapocalyptic" provide clues. The rhythm of our century seemed predictable. World War I at the second decade; World War II at the fourth decade; World War III at the sixth decade, during which the world as we know it comes to an end. But somehow it did not happen when it was supposed to. By the ninth decade, we cannot help suspecting that maybe it happened after all and we failed to notice. Consequently time splits into a false future in which we all live and a true future that by virtue of being true does not have us in it. Variations of this scenario have persisted in our literature and film for years. As early as 1962, Philip K. Dick was writing it in *The Man in the High Castle*; Nabokov pursued a variation in *Ada or Ardor: A Family Chronicle* (1969); Walker Percy played with it in *Love in the Ruins* (1971). More recently, such films as *Back to the Future, Brazil, Terminator*, and *Peggy Sue Got Married* also toy with the idea.

Reinforcing the sense that the future has been used up before it arrives is the acceleration that Alvin Toffler, in a phrase that now seems antiquated, termed future shock. Pop culture changes so quickly that to be current one must be futuristic, for if one is merely up to date one is already out of fashion. Technology seems to change at least as fast, especially information technology. For example, it is estimated that the limits of silicon technology will be reached in about a decade. Research is already being conducted on materials that have the potential to use photons rather than electrons to carry information (gallium arsenide is currently the material of choice). With photonic materials messages travel at the speed of light; consequently transmission times will be even shorter than they now are. In a review of information technology materials for *Scientific American*, John S. Mayo significantly does not conclude that limits inherent in photonic materials will finally act as a brake on development. Rather, he foresees that humans will become increasingly peripheral to information processing. "The higher the functionality of a system is," Mayo writes, "the more important its software component. To reach the highest levels of functionality, [we must] *build systems that are self-directing and independent of intelligent human beings* [emphasis added].[6]

Part of our sense that time has flattened out derives from uncertainty about where we as human beings fit into our own future scenarios. A Saturday morning spent watching cartoons will convince the skeptical how prevalent are our images of ourselves as mere cores on which to encrust cybernetic mechanisms (a theme devel-

[6]John S. Mayo, "Materials for Information and Communiction," *Scientific American* 255 (October 1986):65.

oped at length in *Robocop*).[7] These images imply something more than the usual doubt among the young that history in general and the older generation in particular have anything useful to teach them. They point to a feeling that time itself has ceased to be a useful concept around which to organize experience. If we may or may not be simulacra displaced into the present from previous contexts; if we cannot envision ourselves in the future without imagining that it has undergone a phase change into a different kind of space; if the future is already used up before we can achieve such a projection, problematic as it may be—if all this is true, then time is not just denatured. It is obsolete.

The paradoxical notion that time could be obsolete has been hinted for some time now among theorists interested in postmodern aesthetics. The claim usually is phrased more gently (for obvious reasons, given the paradoxes it engenders), but it is nonetheless widespread. As mentioned earlier, Michel Serres (1982a) sees the temporal aesthetic of nineteenth-century realism giving way to a spatial aesthetic focusing on deformations, local turbulence, and continuous but nondifferentiable curves. Fredric Jameson (1984) has argued that, along with a reconfigured space, a weakening of a sense of historicity is one of the constituent features of postmodernism. Analyzing postmodernism, then, amounts to writing the history of no history. In an important sense, to write the history of postmodernism is to indulge in anachronism.

In this anachronistic history, a dichotomy emerges between those who write about postmodernism and those who live it. Such theorists as Jameson, Andreas Huyssen, Jürgen Habermas, Jean-François Lyotard, Craig Owens, and Hal Foster are concerned to locate postmodernism as part of a historical sequence that began with modernism.[8] They vary in their estimations of how successful, and even

[7]The prevalence of cybernetic imagery in children's toys and television shows is documented by Gabriele Schwab, "Cyborgs, Postmodern Phantasms of Body and Mind," *Discourse* 9 (Spring–Summer 1987): 64–84.

[8]Andreas Huyssen, "Mapping the Postmodern," *New German Critique* 33 (1984): 5–52; Jürgen Habermas, "Modernity versus Postmodernity," *New German Critique* 22 (1981): 3–14, and "Modernity—An Incomplete Project," in Foster, 1983:3–15; Craig Owens, "The Allegorical Impulse: Toward a Theory of Postmodernism, Part 2," *October* 13 (1980): 59–80; Hal Foster, "Postmodernism: A Preface," in Foster, 1983:ix–xvi.

what, modernism was; they disagree about whether it is continued or refuted by postmodernism. Whatever their stance, however, they concur that postmodernism *has a history*, and thus that it has roots in such intellectual issues as the self-referentiality of symbol systems, the Kantian sublime, the cultural logic of late capitalism, and so forth.

The case is very different for those who live postmodernism. For them, the denaturing of time means that they have no history. To live postmodernism is to live as schizophrenics are said to do, in a world of disconnected present moments that jostle one another but never form a continuous (much less logical) progression. The prior experiences of older people act as anchors that keep them from fully entering the postmodern stream of spliced contexts and discontinuous time. Young people, lacking these anchors and immersed in TV, are in a better position to know from direct experience what it is to have no sense of history, to live in a world of simulacra, to see the human form as provisional.[9] The case could be made that the people in this country who know the most about how postmodernism *feels* (as distinct from how to envision or analyze it) are all under the age of sixteen.

Denaturing the Human

Faced with the choice of giving way to the postmodern current or trying to resist it, many adults experience ambivalence. They are uneasy with good reason, for the logic of postmodernism seems inevitably to lead to a fourth wave: the denaturing of the human. Donna Haraway's "Manifesto for Cyborgs" (1985) explores the complexities and nuances of this fourth wave. Haraway points out that information technologies are bringing about far-reaching changes in the way boundaries are conceived and constituted. "Any objects or persons," Haraway writes, "can be reasonably thought of in terms of disassembly and reassembly; no 'natural' architectures constrain system design. The financial districts in all the world's cities, as well

[9]The effect of television has of course extensive implications. Some of them are explored in Todd Gitlin, ed., *Watching Television* (New York: Random House, 1987).

as the export-processing and free-trade zones, proclaim this elementary fact of 'late capitalism'" (p. 81). Haraway argues that information technology has made it possible for us to think of entities (including human beings) as conglomerations that can be taken apart, combined with new elements, and put together again in ways that violate traditional boundaries.

From one perspective this violation is liberating, for it allows historically oppressive constructs to be deconstructed and replaced by new kinds of entities more open to the expression of difference. The problem, of course, is that these new constructs may also be oppressive, albeit in different ways. For example, much feminist thought and writing since the 1940s has been directed toward deconstructing the idea of "man" as a norm by which human experience can be judged. To achieve this goal, another construction has been erected, "woman." Yet as it has been defined in the writings of white, affluent, heterosexual, Western women, this construct has tended to exclude the experiences of black women, Third World women, poor women, lesbian women, and so on. The problem is another form of the local/global problematic, for the new globalized constructs replicate the old in the sense that they too repress local variations in what they purport to describe.

How to create the global forms necessary for effective collective action without repressing difference? Following Chela Sandoval, Haraway suggests that it may be possible to "construct a postmodernist identity out of otherness and difference" through "oppositional consciousness" (p. 73). The idea derives from Sandoval's observation that there are no accepted criteria for determining who qualifies as a woman of color. As a group, women have historically been defined negatively, as not-man. People of color have similarly been defined as not-white. Formed by a double negation, women of color have no positive identity. To Sandoval and Haraway this is a great strength, for it opens the possibility of an identity that operates through negativity rather than positivity. Like Jameson and Gibson, they sense the arrival of a new kind of space. But unlike Jameson and Gibson, they are concerned *not* to define this space, to leave it unconstructed. This of course is self-contradictory, for if the space is unarticulated and unconstructed, how could they write about it and how could we know it exists? The alternative is to fill it with constructions formed by a new kind of dynamic.

Cyborgs as Denatured Humans

The sign Haraway erects to preside over her new kind of space is the cyborg. Essential to her vision of the cyborg as a liberating feminist construct is its violation of traditional boundaries. According to Haraway, at least three distinct oppositions are undone by the cyborg: human/animal ("Far from signaling a walling off of people from other living beings, cyborgs signal disturbingly and pleasurably tight coupling," p. 68); human/machine ("Our machines are disturbingly lively, and we ourselves frighteningly inert," p. 69); and physical/nonphysical ("Our best machines are made of sunshine; they are all light and clean because they are nothing but signals, electromagnetic waves, a section of a spectrum. . . . People are nowhere near so fluid, being both material and opaque. Cyborgs are ether, quintessence," p. 70). The quotations illustrate Haraway's ironic tone as well as her argument. Indeed, her argument could hardly be made without this tone. Irony usually functions to force us to recognize the gap between the ideal and the actual. In Haraway, the ironic tone is exaggerated until the disparity between ideality and actuality becomes, as Samuel Johnson said of *Cymbeline*, a fault too gross for detection.

The tone flaunts what Haraway could scarcely conceal if she tried: that the cyborg, *as an actuality*, is deeply bound up with what Haraway calls the "informatics of domination." The hope that the cyborg can keep the unconstructed space of postmodernism from being filled with newly oppressive constructions is just that—a hope. The reality is that the cyborg has already been appropriated by multinational corporations as they proceed to implement more de-skilling of human labor, more interlocking of data networks, more development of space weapons, and more uncontrollable defense systems.

The point I wish to underscore here is not Haraway's ironic claim that the cyborg can be liberating but her vision of the postmodern as the posthuman. The cyborg is the specific instance for which the denaturing process is the general case. Like the cyborg, denaturing opens up the possibility of unconstructed spaces (although "space" is too narrow a term for what I mean here, since space itself is reconfigured in the postmodern dialectic). And like the cyborg, the denaturing process arouses intense ambivalence, especially as it spreads to envelop the human. Language, context, and time are essential components of human experience. However, the human is a construction logically prior to all three, for it defines the grounds of experience itself. If denaturing the human can sweep away more of the detritus of the past than any of the other postmodern deconstructions, it can also remove taboos and safeguards that are stays, however fragile, against the destruction of the human race. What will happen to the movement for human rights when the human is regarded as a construction like any other? Such concerns illustrate why, at the heart of virtually all postmodernisms, one finds a divided impulse.

Looping from Cultural Postmodernism to Theoretical Postmodernisms

I turn now from talk of postmodernism as a movement determined (and overdetermined) by multiple currents within the culture to discuss specific theories that are postmodern in some sense. With this turn the self-reflexive nature of my argument becomes explicit, for postmodern theories are themselves part of the cultural matrix that I have described as engendering them. Depicting cultural postmodernism as a linear chain of events tends to obscure the complex feedback loops that connect theories to new technologies and social formations, and technologies and social formations to new theories. Narrative linearity, in other words, is fundamentally at odds both with postmodern theories and with the recursiveness of the theory–technology–social-matrix loop. I shall return to this point near the end of the chapter, when I speculate about the relation of cultural postmodernism to narrative structures in general and to my narrative in particular.

Let me begin the present section (which is actually not a new discussion but another recursive loop through the same discussion) by stating my primary premise. *Issues become energized in theories because they are replicated from and reproduced in the social.* In the case of theoretical postmodernisms, this means that at least one of the four constituent waves of postmodernism as a cultural dominant is involved in a feedback loop with the theory. Usually theories that are perceived as postmodern are in a loop through two or more components of cultural postmodernism. Deconstruction, for example, loops recursively between denatured language and denatured context. The new historicism locates itself at the looping of denatured context with denatured time. Jameson's theory of postmodernism, perhaps the single most influential work on the subject to date, is a triple looping through denatured language, context, and time. Haraway identifies the denatured human through the medium of denatured language; and so forth.

To understand how a specific postmodern theory engages in a feedback loop with postmodernism as a cultural dominant, consider the new historicism as a case in point. Edward Pechter has pointed out that the new historicists (in particular Stephen Greenblatt, Alan Sinfield, Jonathan Dollimore, and the early Foucault) all tend to tell the same story. It is the story of how, as Greenblatt puts it, "in all my text and documents, there were, so far as I could tell, no moments of pure, unfettered subjectivity; indeed, the human subject itself began to seem remarkably unfree, the ideological product of the relations of power in a particular society" (quoted in Pechter, 1987:300). Pechter convincingly demonstrates that underlying this story is a "conception of the text as threat, a hostile otherness designed to dominate the reader." Why, he asks, should texts be conceived in this way?

This is the right question to ask. Unfortunately, Pechter does not answer it (presumably because he has another agenda to follow, which is to deny that "human activity is essentially determined by the will to power," p. 301). Before he goes off on this tack, however, he remarks that the new historicist view of texts replicates the "antagonistic power relations at the center of the new-historicist thematics" (p. 300). Granted that the thematics of new historicism replicates its view of texts, what in the present cultural moment makes the story of disempowerment seem compelling?

My answer is that the new historicism is one site within the culture at which the ambivalence characteristic of postmodernism is being played out. The potential liberation that the denaturing process promises is realized in the critic's feeling that she has penetrated the text's defenses and exposed the underlying ideology at work.

The uneasiness to which the denaturing process gives rise is expressed in the kind of stories that are told—stories in which, as Pechter elegantly puts it, "we are only what we are constituted to be by the power relations that govern, anonymously and without human face, even the governors" (p. 300).

The irony is that the more the liberating play of the intellect works on its material, the more power relations are exposed, and consequently the more helpless the subject is made to feel. One could imagine a feedback loop in which the critic writes in order to feel empowered; but the result of her writing is to identify her as a deconstructed subject whose "self" is only the inscription of anonymous forces. Realizing this, she returns to her writing, for now she needs more desperately than ever the feeling of empowerment that comes from writing. But her writing only tells the same story in other ways, and the only relief from it is to write some more. . . . Of course this particular feedback loop probably never develops in the simplified way I have suggested. Intervening are all sorts of other factors, such as the fame and recognition that becoming a successful new historicist brings with it. But the idealized scenario helps to explain how feedback loops develop between the act of theorizing and the thematics of a theory, as well as between postmodernism as a cultural dominant and a specific theory that reinforces and reenacts its concerns. No wonder that a writer such as de Man, who positions himself in opposition to mastery, can seem so masterful. The recursiveness of the culture-theory-technology loop makes this kind of ambiguity all but unavoidable.

How do these loops activate suspicion of global theories, and correlatively, make local knowledge seem so important? The contemporary emphasis on local knowledge is reinforced by the social dialectic Trow addressed when he observed that American life is split between a grid of millions and the circle around the TV set. If there were some way to move between these scales, to coordinate the global and local, then context could be renaturalized. For us at this moment, reinscribing local knowledge in contemporary theories is an important project because it is an attempt to mediate between the denatured contexts of cultural postmodernism. Conversely, globalizing theories seem pernicious because they assume that a single, unified context still exists. When denatured contexts are all around us, we can no longer believe in the innocence of this assumption. But even the project of recontextualizing comes under suspicion when we realize that our constructions grow out of our culture and that our culture is comprised of our constructions, for then it is no longer possible to separate unambiguously the project of reinscribing local knowledge from a globalizing imperative.

Locating the New Scientific Theories in the Postmodern Loop

As we saw in chapter 8, Lyotard in *The Postmodern Condition* (1984) creates something he calls "paralogy" by lumping together quantum mechanics, Gödel's theorem, irreversible thermodynamics, and fractal geometry. In Lyotard's view, paralogy is a new kind of science that will "let us wage a war on totality" (p. 82). The common element in these very different theories, according to Lyotard, is their use of paradox to invalidate global concepts and valorize local knowledge. In chapter 8 I argued that Lyotard's characterization of these theories is misguided, for they have not by any means abandoned globalizing concepts. Nevertheless, he is correct in positing a connection between the new scale-dependent scientific paradigms and the social matrix of postmodernism. This essential insight can be made more useful if we distinguish more carefully between different scientific sites and their various reinscriptions of postmodernism as a cultural dominant.

Consider how space is conceived in fractal geometry. As we have seen, fractal spaces are not Cartesian grids but complex forms characterized by multiple or infinite levels of self-similarity. One can imagine strolling through a city neighborhood along a grid of regularly spaced streets. We know very well how to orient ourselves in such spaces and consider them predictably well ordered. City space reproduces the Cartesian grid, just as the Cartesian grid replicates our experience of city space. By contrast, we do not know very well how to orient ourselves within the complex natural forms represented by mountains, clouds, and galaxies. The usual way people map such spaces is to impose Cartesian grids on them, such as the four compass points. Fractal geometry suggests a different method of orientation, for it sees these spaces as self-similar shapes generated through recursive iterations. Clearly this view does not correlate very well with the scale-invariant properties of city streets and Cartesian maps. What does it correlate with?

Think again of the Bonaventure's lobby—the four identical pillars that mock compass orientation by making it virtually impossible to tell one direction from another; the hanging streamers that convert the rectilinear volume into an irregular, constantly evolving shape. Think of Baudrillard's precession of simulacra, where there are only copies of copies in an endless display of self-similar forms, none of which can be privileged as more "original" than any other. Think of Gibson's cyberspace, in which people and computers become equally sentient entities vying for control of a space that is very real but entirely different from everyday life. I submit that the same forces that authorize these visions have also authorized fractal geometry. Fractal geometry is emerging as an important area of research because it is one way of conceptualizing and understanding postmodern space.

It is also becoming a *source* of this space. In a splendid example of the feedback loop between theory and culture, IBM recently sponsored a series of commercials featuring fractal geometry and Mandelbrot. Fractal geometry is also becoming a commodity in other ways. For example, Heinz-Otto Peitgen and P. H. Richter have used fractal images to market a handsome coffee-table book called *The Beauty of Fractals* (1986). Also available are fractal prints, fractal calendars, fractal T-shirts, and even fractal coffee mugs (which should of course be used on the coffee table with the book). Already a simulation, the computer-generated fractal image is thus transmuted into a series of simulacra in a process that Baudrillard would find predictable. In this regard, it is interesting to note that Mandelbrot has come in for criticism from other applied mathematicians, especially Michael Barnsley, for advocating that chance should continue to play a role in the generation of fractal images.[10] In an essay revealingly titled "Making Dynamical Systems to Order," Barnsley argues that relying on chance makes fractals less useful for modeling

[10]Benoit B. Mandelbrot, "Fractals and the Rebirth of Iteration Theory," in Peitgen and Richter, 1986:151–160; and Michael F. Barnsley, "Making Dynamical Systems to Order," in Barnsley and Demko, 1986:53–68.

than using deterministic procedures that allow for greater control over the image (Barnsley and Demko, 1986:53–68). In view of how quickly fractal geometry has undergone commodification, the outcome of the debate is not difficult to foresee.

A similar correlation between culture and theory obtains between new scientific treatments of time. Time is rarely represented these days by pointers moving across a mechanical clock's face; instead it is signified by the blinking display of an electronic counter. Similarly, time in fractal geometry is not treated as the advancement of points along a number line. Rather, it is conceptualized as small changes in the iterative formulae that are used to generate fractal shapes. We saw in chapter 6 that small changes in these formulae result in large-scale changes in the fractal forms. Consequently, complex shape changes can be described in many fewer bits of information than would be required if one conceived of the shapes as masses of points that had to be advanced through time individually. Thus time still exists in fractal geometry, just as it does in cultural postmodernism. But it is no longer analogous to human movement through a Cartesian plane. Instead, it is envisioned in terms of what one needs to feed into the computer.

The close tie between computer technology and fractal geometry marks it as a site especially responsive to postmodernism as a cultural dominant. Mandelbrot and others have remarked that fractal geometry depends upon computers as no other geometry has done before it. It could equally well be said that computer use has developed in new ways because of fractal geometry. As we saw in chapter 6, computers have allowed mathematics to be practiced as an *experimental* science. When iterating fractal forms, for example, the computer operator does not need to know exactly what will happen. Instead she can feed in the initial values, watch the display, and adjust the variables accordingly as the display continues to change. Computer-generated fractals thus evolve in a continuous fluid interaction between human and machine intelligence. Global form and local inscription merge in this joining of human with computer.

Although I have chosen fractal geometry as an example, the same kinds of consideration apply to chaos theory generally. To show the relation between chaos theory and postmodernism, I will loop one more time through the narrative of postmodernism as I see it. In the early years of this century, efforts were made in a variety of areas to construct unified field theories that would eliminate ambiguity and self-reference. These efforts failed; but the failures brought to light certain intrinsic limits to representation. Having swung as far as it could in the direction of closure, the pendulum began to swing the other way as people became interested in exploring the implications of ungrounded representation. At the same time that global networks of communications, finances, energy sources, weapons research, and so forth made human lives on the planet more interdependent than they had ever been before, theoretical postmodernisms put forth urgent claims for fragmentation, discontinuities, and local differences. Cultural postmodernism arises from loops that build up between these seemingly divergent trends.

In its theoretical guises, cultural postmodernism champions the disruption of globalized forms and rationalized structures. In its technological guises, it continues to erect networks of increasing scope and power. Despite their apparent opposition, these two aspects of cultural postmodernism engage each other in self-sustaining feedback loops. Anxieties about the networks are expressed by theoretical postmodernisms; insofar as these expressions are acts of empowerment, however, they also relieve anxiety—and the networks continue to expand. It is the same paradox we saw in de Man, where the critique of mastery became so entwined with mastery that the two were inextricable. It is the same story told by new historicists, who narrate it even while it inscribes itself in the dialectic between them and their writing.

The science of chaos shares with other postmodernisms a deeply ingrained ambivalence toward totalizing structures. On the one hand, it celebrates the disorder that earlier scientists ignored or disdained, seeing turbulent flow not as an obstacle to scientific progress but as a great swirling river of information that rescues the world from sterile repetition. On the other hand, it also shows that when one focuses on the underlying recursive symmetries, the deep structures underlying chaos can be revealed and analytical solutions can sometimes be achieved. It is thus like other postmodernisms in that it both resists and contributes to globalizing structures.

It differs in its particular expression of this ambivalence. Chaos theorists do not perceive themselves as creating chaos, either through their discourse or through their representations. They believe that they merely recognize what was always there but not much noticed before. In other postmodernisms, it is not so clear that chaos inheres in the system rather than the theorist. In deconstruction, for example, the "always already" formula asserts that texts were *always* chaotic before anyone wrote about them, certainly before deconstruction appeared on the scene. But since texts were not generally perceived to be chaotic before deconstruction, the burden of proof is on the writer to show that it is not she who creates the chaos she reveals. This necessity leads to a certain amount of anxiety, for what the writer bestows the writer can take away. Deconstruction is therefore concerned that the disorder it celebrates and exposes can be undone if one is not unceasingly vigilant about how one writes and speaks.

I emphasize this difference between chaos theory and deconstruction because I believe that it is important to distinguish as carefully as possible between one site and another. The general patterns may be (recursively) the same, but the tonalities and precise dynamics of each site are unique. To speak of the sciences of chaos as postmodern science is not in my view to speak incorrectly. It is to speak carelessly, however, unless one specifies the tensions that mark a specific site and re-mark it with the distinctive dynamics that characterize it. Fractal geometry is not the same as nonlinear dynamics; nonlinear dynamics is not the same as information theory; information theory is not the same as thermodynamics. It goes without saying that greater differences distinguish any of these fields of inquiry from deconstruction and from the new historicism. While it is important to see that all postmodern theories have certain common characteristics that reinforce and are reinforced by the culture, it is also important to understand that different sites impart distinctive values to the ideas they share.

Having acknowledged the importance of local differences, I should like to make a global conjecture about why the sciences of chaos have been so energized by cultural postmodernism. Many scientists have commented that working on chaos has allowed them to renew their sense of wonder.[11] Although they do not put it this way,

[11]This aspect of chaos theory is eloquently addressed in James Gleick's *Chaos* (1987).

they intimate that chaos has given them a sense of being in touch with the Lacanian real. For them, chaos is an image for what can be touched but not grasped, felt but not seen. At a time when resistance to mastery is so sophisticated that it cannot help but be perceived as masterful, chaos presents them with a resistance that alleviates the fear of mastery. They see it as rescuing postmodernism from the prefix that positions us in "always already." They take chaos as demonstrating that there is something more than novelty, something other than the precession of simulacra. For them, chaos signifies the truly new. This is what they believe.

When chaos theory is located within the postmodern dialectic, however, a more complex picture emerges. As we saw in chapter 6, the same structures that have given us a gendered world also operate in the construction of chaos. The innocence of chaos is an assumption that is most tenable when one believes that the self is not itself constructed by the same forces that are replicated in theories and in the social matrix. When theory, self, and culture are caught in the postmodern loop, the construction of chaos cannot be unambiguous, because it derives from and feeds into the same forces that made us long for escape. Like all postmodern theories, chaos theory is marked by the ambivalence characteristic of this cultural moment.

The Story of Chaos: Denaturing Narratives

This, then, is the story of chaos and culture as I see it. It remains to say what status I wish to claim for my narrative. How can one write about postmodernism without being acutely aware that what one writes is itself a construction in the postmodern loop? For my part, I am conscious that out of the nearly infinite number of events that have happened in this century, I have chosen some few dozen to remark upon and weave together in a pattern. There can be no question that by choosing different locales, one could weave entirely different or contradictory patterns. Moreover, the assumptions informing my narrative are no doubt full of the very contradictions and ambivalences that I have described as characteristic of postmodernism. No, I have not described the social matrix. Even if my account were fuller and more complete than it is, I could not have achieved the vantage outside or above my culture which an "objective" description would imply. At best I have reenacted the cultural dominant in such a way as to make its dynamics clearer than they may have been before. If my narrative is useful, it is because it self-consciously embraces what it cannot help being—a denatured construction.

Assuming that cultural history is always a dialogue between critics rather than an objective record of what happened, one would suppose that this dialogue changes over time, and that this evolution itself constitutes a narrative, or more properly a metanarrative. When Lyotard defined postmodernism in *The Postmodern Condition* as "incredulity toward metanarratives," he had in mind such specific social narratives as the story of scientific progress and the rise of democratic education (p. xxiv). But what are the essential components of narrative construction, if not language, context, time, and the human? The denaturing of experience, in other words, constitutes a cultural metanarrative; and its peculiar property is to imply incredulity not just toward other metanarratives but toward narrative as a form of representation. It thus implies its own deconstruction.

The denaturing of experience has important implications for narrative literature, especially narrative fiction. The denaturing process can be thought of as a set of vectors defining a textual field of play. The more vectors that are energized in a particular text or site, the more dimensional complexity the field will have, and consequently the more complicated it can be. In a fully denatured narrative, one would expect the language to be self-referential; the context to be self-consciously created, perhaps by the splicing together of disparate contexts; the narrative progression to be advanced through the evolution of underlying structures rather than through chronological time; and the characters to be constructed so as to expose their nature as constructions.

However, this list of literary strategies should not be construed as a set of criteria by which one can determine which narratives are "really" denatured or postmodern. Such a project would be self-contradictory, for it presupposes that the criteria exist in an ideal reality removed from the phenomena they attempt to describe, whereas in fact they derive from a theory that is itself frankly a construction involved in a recursive loop with postmodern culture.

Although I have defined cultural postmodernism as the denaturing of experience and have placed it within the time frame of the twentieth century, the literary strategies mentioned above can be found in texts from virtually any period. What could be more self-referential than the end of *A Midsummer Night's Dream*, or more effective at representing the denatured human than *Frankenstein*? Postmodern texts do not have a monopoly on these literary strategies. It is not the literary strategies in isolation that make a text postmodern but rather their connection through complex feedback loops with postmodernism as a cultural dominant. Other times have had glimpses of what it would mean to live in a denatured world. But never before have such strong feedback loops among culture, theory, and technology brought it so close to being a reality.

Selected Bibliography

SCIENTIFIC SOURCES

Atlan, Henri. 1974. "On a Formal Definition of Organization." *Journal of Theoretical Biology* 45:295–304.

Barnsley, Michael F., and Stephen G. Demko, eds. 1986. *Chaotic Dynamics and Fractals*. Orlando: Academic Press.

——, V. Ervin, D. Hardin, and J. Lancaster. 1986. "Solution of an Inverse Problem for Fractals and Other Sets." *Proceedings of the National Academy of Sciences*, Mathematics, 83, no. 7 (April): 1975–1977.

Bennett, Charles H. 1987. "Demons, Engines, and the Second Law." *Scientific American* 258 (November): 108–116.

—— and Rolf Landauer. 1985. "The Fundamental Physical Limits of Computation." *Scientific American* 253 (July): 48–71.

Boltzmann, Ludwig. 1909. *Wissenschaftliche Abhandlungen von Ludwig Boltzmann*. Ed. Fritz Havenohrl. 3 vols. Leipzig: Barth.

Brillouin, Leon. 1951. "Maxwell's Demon Cannot Operate: Information and Entropy. I." *Journal of Applied Physics* 22 (March): 334–357.

——. 1956. *Science and Information Theory*. New York: Academic Press. 2d ed. 1962.

Cardwell, D. S. L. 1971. *From Watt to Clausius: The Rise of Thermodynamics in the Early Industrial Age*. Ithaca: Cornell University Press.

Chaitin, Gregory J. 1975. "Randomness and Mathematical Proof." *Scientific American* 232 (May): 47–52.

Clausius, R. J. E. 1850. "On the Motive Power of Heat, and on the Laws Which Can Be Deduced from It for the Theory of Heat." In *Annalen der Physik*, ed. J. C. Poggendorf, 84:368–500.

Cohen, D. S., J. C. Neu, and R. R. Rosales. 1978. "Rotating Spiral Wave Solutions to Reaction-Diffusion Equations." *SIAM Journal of Applied Mathematics* 35:536–547.

Coveney, Peter V. 1988. "The Second Law of Thermodynamics: Entropy, Irreversibility, and Dynamics." *Nature* 333:409–415.

Crutchfield, James P., J. Doyne Farmer, Norman H. Packard, and Robert S. Shaw. 1986. "Chaos." *Scientific American* 255 (December): 46–57.

Davies, Paul. 1984. *Superforce: The Search for a Grand Unified Theory of Nature*. New York: Simon & Schuster.

Duffy, M. R., N. F. Britton, and J. D. Murray. 1980. "Spiral Wave Solutions of Practical Reaction-Diffusion Systems." *SIAM Journal of Applied Mathematics* 39:8–12.

Einstein, Albert. 1954. *Ideas and Opinions*. New York: Crown.

Feigenbaum, Mitchell J. 1980. "Universal Behavior in Nonlinear Systems." *Los Alamos Science* 1 (Summer): 4–27.

Feynman, Richard P. 1986. *QED: The Strange Theory of Light and Matter*. Princeton: Princeton University Press.

Ford, Joseph. 1983. "How Random Is a Coin Toss?" *Physics Today* 36 (April): 40–47.

——. 1986. "Chaos: Solving the Unsolvable, Predicting the Unpredictable!" In *Chaotic Dynamics and Fractals*, ed. Michael F. Barnsley and Stephen G. Demko, pp. 1–52. New York: Academic Press.

Gardner, Martin. 1979. "Mathematical Games." *Scientific American* 241 (November): 20–34.

Gleick, James. 1987. *Chaos: Making a New Science*. New York: Viking.

Gödel, Kurt. 1962. *On Formally Undecidable Propositions in "Principia Mathematica" and Related Systems*. Ed. R. B. Braitewaite. Trans. Bernard Meltzer. New York: Basic Books. Trans. of "Über formal unentscheidbare Sätze der Principia Mathematica unter verwandte Systeme I." *Monatshefte für Mathematik und Physik* 38 (1931): 173–198.

Gunzig, Edgard, Jules Geheniau, and Ilya Prigogine. 1987. "Entropy and Cosmology." *Nature* 330:621–624.

Hawking, Stephen. 1988. *A Brief History of Time: From the Big Bang to Black Holes*. New York: Bantam.

Hudson, J. L., and C. Mankin. 1981. "Chaos in the Belousov-Zhabotinskii Reaction." *Journal of Chemical Physics* 74:6171–6177.

Jacob, François. 1976. *The Logic of Life: A History of Heredity*. Trans. Betty E. Spillmann. New York: Vintage.

Jorna, Siebe, ed. 1978. *Topics in Nonlinear Dynamics*, vol. 46 of *AIP Conference Proceedings*. New York: American Institute of Physics.

Kelvin, Lord (William Thomson). 1881; 1911. *Mathematical and Physical Papers*. Vols. 1 and 5. Cambridge: Cambridge University Press.

Kuhn, Thomas. 1970. *The Structure of Scientific Revolutions*. 2d ed. Chicago: University of Chicago Press.

Lorenz, Edward. 1963. "Deterministic Nonperiodic Flow." *Journal of the Atmospheric Sciences* 20:130–141.

MacKay, Donald M. 1969. *Information, Mechanism, and Meaning*. Cambridge: MIT Press.

Madore, Barry F., and Wendy L. Freedman. 1987. "Self-organizing Structures." *American Scientist* 75:252–259.

Mandelbrot, Benoit B. 1983. *The Fractal Geometry of Nature*. New York: W. H. Freeman.

Maxwell, James Clerk. 1860. "Illustrations of the Dynamical Theory of Gases." *Philosophical Magazine* 19 (January): 19–32; 20 (July): 21–37.

——. 1871. *Theory of Heat*. London and New York: Longmans, Green.

——. 1890. *Scientific Papers of James Clerk Maxwell*. Ed. W. D. Niven. 2 vols. Cambridge: Cambridge University Press. Reprinted New York: Dover, 1952.

Monod, Jacques. 1972. *Chance and Necessity: An Essay on the Natural Philosophy of Modern Biology*. Trans. Austryn Wainhouse. New York: Vintage.

Mueller, M. W., and W. D. Arnett. 1976. "Propagating Star Formation and Irregular Structure in Spiral Galaxies." *Astrophysics Journal* 210:670–678.

Newall, P. C., and F. M. Ross. 1982. "Inhibition by Adenosine of Aggregation Centre Inhibition and Cyclic AMP Binding in Dictyostelium." *Journal of General Microbiology* 128:2715–2724.

Nicolis, G., and I. Prigogine. 1977. *Self-Organization in Nonequilibrium Systems: From Dissipative Structures to Order through Fluctuations*. New York: Wiley.

Parker, Barry. 1986. *Einstein's Dream: The Search for a Unified Theory of the Universe*. New York: Plenum.

Peitgen, Heinz-Otto, and P. H. Richter, eds. 1986. *The Beauty of Fractals: Images of Complex Dynamical Systems*. Berlin and New York: Springer.

Peterson, Ivars. 1987. "Packing It In: Fractals Play an Important Role in Image Compression." *Science News* 131 (May 2): 283–285.

Poincaré, Jules Henri. 1890. "Sur le problème des trois corps et les équations de la dynamique," *Acta Mathematica* 13:1–270.

Popper, Karl L. 1965. *Conjectures and Refutations: The Growth of Scientific Knowledge*. 2d ed. New York: Basic Books.

Prigogine, Ilya, and Isabelle Stengers. 1984. *Order out of Chaos: Man's New Dialogue with Nature*. New York: Bantam.

Roux, J. C., A. Rossi, S. Bachelart, and C. Vidal. 1980. "Representation of a Strange Attractor from an Experimental Study of Chemical Turbulence." *Physics Letters* 77A:391–393.

Shannon, Claude E. 1948. "A Mathematical Theory of Information." *Bell System Technical Journal* 27 (July and October): 379–423, 623–656.

——. 1951. "Prediction and Entropy of Printed English." *Bell System Technical Journal* 30:50–64.

—— and Warren Weaver. 1949. *The Mathematical Theory of Communication*. Urbana: University of Illinois Press.

Shaw, Robert. 1981. "Strange Attractors, Chaotic Behavior, and Information Flow." *Zeitschrift für Naturforschung* 36A (January): 79–112.

Simoyi, Reuben, Alan Wolf, and Harry L. Swinney. 1982. "One-Dimensional Dynamics in a Multicomponent Chemical Reaction." *Physical Review Letters* 49:245–248.

Singh, Jagjit. 1966. *Great Ideas in Information Theory, Language and Cybernetics*. New York: Dover.

Szilard, Leo. 1929. "On the Reduction of Entropy as a Thermodynamic System Caused by Intelligent Beings." *Zeitschrift für Physik* 53:840–856.

Tritton, David. 1986. "Chaos in the Swing of a Pendulum." *New Scientist* 24 (July): 37–40.

Venturi, Robert, Denise Scott Brown, and Steven Izenour. 1972. *Learning from Las Vegas: The Forgotten Symbolism of Architectural Form*. Cambridge: MIT Press. Rev. ed. 1977.

Welsh, B., J. Gomatom, and A. Burgess. 1983. "Three-Dimensional Chemical Waves in the Belousov-Zhabotinskii Reaction." *Nature* 304:611–614.

Whitehead, A. N., and Bertrand A. W. Russell. 1910–1913. *Principia Mathematica*. 3 vols. Cambridge: Cambridge University Press.

Wicken, Jeffrey S. 1987. "Entropy and Information: Suggestions for a Common Language." *Philosophy of Science* 54:176–193.

Wilson, John Arthur. 1968. "Entropy, Not Negentropy." *Nature* 219:535–536.

Wilson, Kenneth G. 1983. "The Renormalization Group and Critical Phenomena." *Reviews of Modern Physics* 55:583–600.

Winfree, A. T. 1980. *The Geometry of Biological Time*. Biomathematics, vol. 8. Berlin and New York: Springer.

Winfree, Arthur T., and S. H. Strogatz. 1985. "Organizing Centers for Three-Dimensional Chemical Waves." *Nature* 311:611–615.

LITERARY SOURCES

Adams, Henry. 1913. *Mont-Saint-Michel and Chartres*. Boston: Houghton Mifflin.

——. 1919. "A Letter to American Teachers of History." In *Degradation of Democratic Dogma*, ed. Brooks Adams. New York: Macmillan.

——. 1938. *Letters of Henry Adams, 1892–1918*. Ed. Worthington Chauncey Ford. Vol. 2. Boston: Houghton Mifflin.

——. 1973. *The Education of Henry Adams*. Ed. Ernest Samuel. Boston: Houghton Mifflin. First published 1907.

Balcerzan, Edward. 1975. "Seeking Only Man: Language and Ethics in *Solaris*." Trans. Konrad Brodzinski. *Science-Fiction Studies* 2:152–156.

Barthes, Roland. 1974. *S/Z*. Trans. Richard Miller. New York: Hill & Wang.

——. 1986. *The Rustle of Language*. Trans. Richard Howard. New York: Hill & Wang.

Baudrillard, Jean. 1981. *For a Critique of the Political Economy of the Sign.* Trans. Charles Levin. St. Louis: Telos Press.

——. 1983. *Simulations.* Trans. Paul Foss, Paul Patton, and Philip Beitchman. Foreign Agent Series. New York: Semiotext(e).

Beagle, Peter S. 1983. "Lem: Science Fiction's Passionate Realist." *New York Times Book Review*, March 20, pp. 7, 33.

Beer, Gillian. 1983. *Darwin's Plots: Evolutionary Narratives in Darwin, George Eliot, and Nineteenth-Century Fiction.* London: Routledge & Kegan Paul.

Benjamin, Walter. 1969. *Illuminations.* Trans. Harry Zohn. New York: Schocken.

Blackmur, R. P. 1980. *Henry Adams.* Ed. Veronica A. Makowsky. New York: Harcourt Brace Jovanovich.

Borges, Jorge Luis. 1964. *Other Inquisitions: 1937–1952.* Trans. Ruth L. C. Sims. Austin: University of Texas Press.

——. 1981. *Borges: A Reader.* Ed. Emir R. Monegal and Alastair Reed. New York: E. P. Dutton.

Calvino, Italo. 1968. *Cosmicomics.* Trans. William Weaver. San Diego: Harcourt Brace Jovanovich.

Campbell, Jeremy. 1982. *Grammatical Man: Information, Entropy, Language, and Life.* New York: Simon & Schuster.

Cox, James M. 1980. "Learning through Ignorance: *The Education of Henry Adams.*" *Sewanee Review* 88:198–227.

Davenport, Marcia. 1954. *My Brother's Keeper.* New York: Scribner.

De Man, Paul. 1979a. *Allegories of Reading: Figural Language in Rousseau, Nietzsche, Rilke, and Proust.* New Haven: Yale University Press.

——. 1979b. "Shelley Disfigured." In *Deconstruction and Criticism*, ed. Harold Bloom et al. pp. 39–74. New York: Seabury.

——. 1982. "The Resistance to Theory." *Yale French Studies* 63:3–20.

Derrida, Jacques. 1968. "La Différance." *Bulletin de la société française de philosophie* 62 (July–September): 73–101.

——. 1976. *Of Grammatology.* Trans. Gayatri C. Spivak. Baltimore: Johns Hopkins University Press.

——. 1977a. "Limited Inc abc . . ." In *Glyph 2: Johns Hopkins Textual Studies*, ed. Samuel Weber and Henry Sussman, pp. 162–254. Baltimore: Johns Hopkins University Press.

——. 1977b. "Signature Event Context." In *Glyph 1: Johns Hopkins Textual Studies*, ed. Samuel Weber and Henry Sussman, pp. 172–197. Baltimore: Johns Hopkins University Press.

Dick, Philip K. 1962. *The Man in the High Castle.* New York: Putnam.

Draine, Betsy. 1980. "Nostalgia and Irony: The Postmodern Order of *The Golden Notebook.*" *Modern Fiction Studies* 26:31–48.

Foster, Hal, ed. 1983. *The Anti-Aesthetic: Essays on Postmodern Culture.* Port Townsend, Wash.: Bay Press.

Foucault, Michel. 1970. *The Order of Things: An Archaeology of the Human Sciences*. New York: Pantheon.

——. 1973. *Madness and Civilization: A History of Insanity in the Age of Reason*. Trans. Richard Howard. New York: Vintage.

——. 1977. *Discipline and Punish: The Birth of the Prison*. Trans. Alan Sheridan. New York: Pantheon.

——. 1980a. *A History of Sexuality*. Vol. 1, *An Introduction*. Trans. Robert Hurley. New York: Random House.

——. 1980b. *Power/Knowledge: Selected Interviews and Other Writings, 1972–1977*. Ed. Colin Gordon. Trans. C. Gordon et al. New York: Pantheon.

Froula, Christine. 1985. "Quantum Physics/Postmodern Metaphysics: The Nature of Jacques Derrida," *Western Humanities Review* 39:287–311.

Geertz, Clifford. 1973. *The Interpretation of Cultures*. New York: Basic Books.

——. 1983. *Local Knowledge: Further Essays in Interpretive Anthropology*. New York: Basic Books.

Gibson, William. 1984. *Neuromancer*. New York: Ace.

Greenland, Colin. 1986. "A Nod to the Apocalypse: An Interview with William Gibson." *Foundation* 36 (Summer): 5–9.

Hague, John A., ed. 1964. *American Character and Culture in a Changing World: Some Twentieth-Century Perspectives*. Westport, Conn.: Greenwood.

Haraway, Donna. 1985. "A Manifesto for Cyborgs: Science, Technology, and Socialist Feminism in the 1980s." *Socialist Review* 80:65–107.

Harbert, Earl N., ed. 1981. *Critical Essays on Henry Adams*. Boston: G. K. Hall.

Hartman, Geoffrey H. 1976. "Monsieur Texte II: Epiphony in Echoland." *Georgia Review* 30 (Spring): 169–188.

Hayles, N. Katherine. 1984. *The Cosmic Web: Scientific Field Models and Literary Strategies in the Twentieth Century*. Ithaca: Cornell University Press.

Heidegger, Martin. 1962. *Being and Time*. Trans. J. Macquarrie and E. S. Robison. New York: Harper & Row.

Hesiod. 1983. *Theogony*. Trans. Apostolos N. Athanassakis. Baltimore: Johns Hopkins University Press.

Hynes, Joseph. 1973. "The Construction of *The Golden Notebook*." *Iowa Review* 4100–113.

Jameson, Fredric. 1984. "Postmodernism, or The Cultural Logic of Late Capitalism." *New Left Review* 146:53–92.

Jarzebski, Jerzy. 1977. "Stanislaw Lem: Rationalist and Visionary. Trans. Franz Rottensteiner. *Science-Fiction Studies* 4:110–126.

Johnson, Barbara. 1980. "BartheS/BalZac." In *The Critical Difference: Essays in the Contemporary Rhetoric of Reading*, pp. 3–12. Baltimore: Johns Hopkins University Press.

Jordy, William. 1952. *Henry Adams: Scientific Historian*. New Haven: Yale University Press.

Kristeva, Julia. 1980. *Desire in Language: A Semiotic Approach to Literature and Art*. Ed. Leon R. Roudiez. Trans. Thomas Gora, Alice Jardine, and Leon R. Roudiez. New York: Columbia University Press.

Landon, Brooks. 1987. "Future So Bright They Gotta Wear Shades." *Cinefantastique* 18 (December): 27–31.

Lem, Stanislaw. 1971a. "Robots in Science Fiction." Trans. Franz Rottensteiner. In *SF: The Other Side of Realism*, ed. Thomas D. Clareson, pp. 307–325. Bowling Green: Popular Press.

——. 1971b. *A Perfect Vacuum: Perfect Reviews of Nonexistent Books*. Trans. Michael Kandel. New York: Harcourt Brace Jovanovich.

——. 1973a. "Culture and Futurology." *Polish Perspectives* (Warsaw) 16:30–38.

——. 1973b. "On the Structural Analysis of Science Fiction." *Science-Fiction Studies* 1:26–33.

——. 1974a. "The Time-Travel Story and Related Matters of SF Structuring." *Science-Fiction Studies* 1:143–154.

——. 1974b. *The Cyberiad: Fables for the Cybernetic Age*. Trans. Michael Kandel. New York: Harcourt Brace Jovanovich.

——. 1981. "Metafantasia: The Possibilities of Science Fiction." Trans. Etelka de Laczay and Istvan Csicsery-Ronay. *Science-Fiction Studies* 8:54–71.

——. 1983. *His Master's Voice*. Trans. Michael Kandel. New York: Harcourt Brace Jovanovich.

——. 1984. "Chance and Order." Trans. Franz Rottensteiner. In *New Yorker* January 30, pp. 88–98.

Lessing, Doris. 1962. *The Golden Notebook*. New York: Bantam.

——. 1971. "Introduction." In *The Golden Notebook*. New York: Bantam.

——. 1974. *A Small Personal Voice: Essays, Reviews, Interviews*. Ed. Paul Schlueter. New York: Knopf.

Levi, Primo. 1985. *If Not Now, When?* Trans. William Weaver. New York: Summit.

Libby, Marion Vlastos. 1974. "Sex and the New Woman in *The Golden Notebook*." *Iowa Review* 5:106–120.

Lyotard, Jean-François. 1984. *The Postmodern Condition: A Report on Knowledge*. Trans. Geoff Bennington and Brian Massumi. Theory and History of Literature, vol. 10. Minneapolis: University of Minnesota Press.

Martin, Ronald. 1981. *American Literature and the Universe of Force*. Durham, N.C.: Duke University Press.

Morgan, Ellen. 1973. "Alienation of the Woman Writer in *The Golden Notebook*." *Contemporary Literature* 14:471–480.

Nabokov, Vladimir. 1969. *Ada or Ardor: A Family Chronicle*. New York: McGraw-Hill.

Occhiogrosso, Frank. 1980. "Threats to Rationalism: John Fowles, Stanislaw Lem, and the Detective Story." *American Detective* 13:4–7.

Pechter, Edward. 1987. "The New Historicism and Its Discontents: Politicizing Renaissance Drama." *PMLA* 103 (May): 292–303.

Peckham, Morse. 1967. *Man's Rage for Chaos: Biology, Behavior, and the Arts*. New York: Schocken.

Philmus, Robert M. 1986. "The Cybernetic Paradigms of Stanislaw Lem." In *Hard Science-Fiction*, ed. George Slusser and Eric Rabkin, pp. 177–213. Carbondale: Southern Illinois University Press.

Poovey, Mary. 1986. "'Scenes of an Indelicate Character': The Medical 'Treatment' of Victorian Women." *Representations* 14 (Spring): 137–168.

Ricoeur, Paul. 1976. *Interpretation Theory: Discourse and the Surplus of Meaning*. Fort Worth: Texas Christian University Press.

Rothfork, John. 1977. "Cybernetics and Humanistic Fiction: Stanislaw Lem's *The Cyberiad*." *Research Studies* 45:123–133.

——. 1981. "Having Everything Is Having Nothing." *Southwest Review* 66:293–306.

Rowe, John Carlos. 1976. *Henry Adams and Henry James: The Emergence of a Modern Consciousness*. Ithaca: Cornell University Press.

Serres, Michel. 1975. *Feux et signaux de brume: Zola*. Paris: Grasset.

——. 1980. *Hermes V: Le Passage du nord-ouest*. Paris: Minuit.

——. 1982a. *Hermes: Literature, Science, Philosophy*. Ed. Josue V. Harari and David F. Bell. Baltimore: Johns Hopkins University Press.

——. 1982b. *The Parasite*. Trans. Lawrence R. Schehr. Baltimore: Johns Hopkins University Press.

Sprague, Claire. 1982. "Doubletalk and Doubles Talk in *The Golden Notebook*." *Papers on Language and Literature* 18:181–197.

Trow, George W. S. 1978. *Within the Context of No Context*. Boston: Little, Brown.

Westervelt, Linda A. 1984. "Henry Adams and the Education of His Readers." *Southern Humanities Review* 18:23–37.

Index

Library of Congress Cataloging-in-Publication Data

Hayles, N. Katherine.
Chaos bound : orderly disorder in contemporary literature and science / N. Katherine Hayles.
p. cm.
Includes bibliographical references.
ISBN 0-8014-2262-0. —ISBN 0-8014-9701-9 (pbk.)
1. Literature, Modern—20th century—History and criticism.
2. Chaotic behavior in systems in literature. I. Title.
PN771.H35 1990
809'.04—dc20 89-23893